Advances in Optical Fiber Communications

Advances in Optical Fiber Communications

Editor

Giuseppe Rizzelli

MDPI • Basel • Beijing • Wuhan • Barcelona • Belgrade • Manchester • Tokyo • Cluj • Tianjin

Editor
Giuseppe Rizzelli
Polytechnic of Turin
Italy

Editorial Office
MDPI
St. Alban-Anlage 66
4052 Basel, Switzerland

This is a reprint of articles from the Special Issue published online in the open access journal *Applied Sciences* (ISSN 2076-3417) (available at: https://www.mdpi.com/journal/applsci/special_issues/Optical_Fiber_Communication).

For citation purposes, cite each article independently as indicated on the article page online and as indicated below:

LastName, A.A.; LastName, B.B.; LastName, C.C. Article Title. *Journal Name* **Year**, *Volume Number*, Page Range.

ISBN 978-3-0365-4947-7 (Hbk)
ISBN 978-3-0365-4948-4 (PDF)

Contents

About the Editor

Giuseppe Rizzelli

Giuseppe Rizzelli is senior researcher at the LINKS Foundation and lecturer at Politecnico di Torino in Turin, Italy. He received his PhD in Electronics Engineering from Alcalá University (Madrid, Spain) in 2018, where he worked on a distributed Raman amplification for both long-haul and unrepeatered optical communications. Prior to enrolling in the PhD program, he was an affiliate researcher at the Lawrence Berkeley National Laboratory (CA, USA) where he worked on mode-locked lasers' synchronization. He recently joined the Computing Photonics and Electromagnetism division of the LINKS Foundation, where his main research focus is on short-reach optical communications for both for metro access and data-center scenarios.

Preface to "Advances in Optical Fiber Communications"

Given the increasing importance of a globally interconnected world, driven by modern digital services and the need for fast and reliable access to digital resources, communications networks are one of the key infrastructures in society at present. Telecom operators strive to make networks increasingly efficient and low-cost at every step of the technology's development. In this scenario, fiber optics and optical devices play a leading role, as they allow for unprecedented growth in our capacity to cope with the ever-increasing traffic demand.

At present, most worldwide communications rely on optical technologies, and most research and industrial efforts focus on achieving the optimum trade-off between transmission speed and cost per bit. This goal is usually pursued by acting on three main aspects: the manufacture of low-cost devices, the introduction of digital solutions to help to overcome some of the physical limitations of optical communications systems and the optimization of network design.

Contributions to this Special Issue address the three aforementioned subjects and provide valuable insights into the optical fiber communications world: Li and coworkers analyze how substrate misorientation affects the structural and optical properties of Quantum Well (QW) lasers with a large lattice mismatch between the InGaAs QW and the GaAs substrate. De Farias and coworkers propose two heuristic strategies to solve the regenerator placement problem (RPP) in translucent networks under dynamic traffic. Hadi and coworkers and Wang and coworkers apply digital predistortion in the radio in fiber and PON scenarios, respectively. Ozolins and coworkers show up to 100 GBaud capacity over 400 m using OOK and PAM-4 modulation and a monolithically integrated, externally modulated laser with 100 GHz 3 dB bandwidth.

The editor would like to thank all the authors for their valuable contributions to the scientific community, and sll the peer-reviewers for taking the time to improve the quality of the submitted manuscripts and, by extension, of the whole Special Issue and editorial team of Applied Sciences for their hard work. My gratitude goes especially to the Section Managing Editor Chloe Liu for her dedication and continued support.

Giuseppe Rizzelli
Editor

applied sciences

Editorial

Advances in Optical Fiber Communications

Giuseppe Rizzelli

Politecnico di Torino, Corso Duca degli Abruzzi 24, 10129 Torino, Italy; giuseppe.rizzelli@polito.it

Citation: Rizzelli, G. Advances in Optical Fiber Communications. *Appl. Sci.* **2022**, *12*, 4818. https://doi.org/10.3390/app12104818

Received: 6 May 2022
Accepted: 9 May 2022
Published: 10 May 2022

1. Introduction

Given the increasing importance of a globally interconnected world, driven by modern digital services and the need for fast and reliable access to digital resources, communications networks are one of the key infrastructures in today's society. Telecom operators strive to make networks increasingly efficient and low-cost at every step of the technology's development. In this scenario, fiber optics and optical devices play a leading role, as they allow for an unprecedent growth in our capacity to cope with the ever-increasing traffic demand.

Currently, most worldwide communications rely on optical technologies [1], and most research and industrial efforts focus on achieving the optimum trade-off between transmission speed and cost per bit. This goal is usually pursued by acting on three main aspects: the manufacture of low-cost devices [2], the introduction of digital solutions to help overcome some of the physical limitations of optical communications systems [3] and the optimization of network design [4].

2. Advances in Optical Fiber Communications

Contributions to this Special Issue address the three aforementioned subjects and bring valuable insights into the optical fiber communications world.

Li and coworkers [5] analyze in detail how substrate misorientation affects the structural and optical properties of Quantum Well (QW) lasers with large lattice mismatch between the InGaAs QW and the GaAs substrate. Understanding how QW lasers work is vital as they are a potential candidate for low-cost semiconductor lasers with high efficiency in uncooled operations [6]. In this manuscript, a mechanism elucidating the effect of substrate miscuts on the structural and optical properties of QWs is proposed and verified.

De Farias and coworkers [7] propose two heuristic strategies to solve the regenerator placement problem (RPP) in translucent networks under dynamic traffic. The translucent architecture is opposed to the fully transparent (all-optical) and fully opaque (optical–electrical–optical regeneration at every intermediate node) network configurations and uses a limited number of regenerators. This is achieved by solving the regenerator placement problem to reduce capital expenditure, while keeping performance at the desired level. The proposed dynamic load distribution (DLD) algorithm outperforms previously proposed heuristics, yielding lower blocking probability for the same number of regenerators, regardless of the topology, traffic pattern or intensity, and considered transmission reach.

The digital aid is introduced by Hadi and coworkers [8] and by Wang and coworkers [9]. In the former paper, digital predistortion is applied in the radio over fiber (RoF) scenario, using either conventional Volterra-based or deep learning methods to compensate for nonlinear impairments in optical front haul-based RoF systems. Both proposed methods which reduce the adjacent channel power ratio and the error vector magnitude. A considerable complexity reduction was also achieved by using a novel magnitude selective affine (MSA) method for digital predistortion.

Wang and coworkers [9] apply digital pre-compensation to the passive optical network scenario, showing a 29 dB power budget in a 200 Gbps communication system based on

PAM-8 modulation over 20 km of standard single mode fiber. Pre-distortion at the transmitter is combined with three different digital signal processing techniques, with the receiver analyzing transmission performance as a function of overall digital complexity. A Volterra nonlinear equalization combined with a square-root-like technique was demonstrated to be the best among the considered options.

Lastly, in even-shorter-reach scenarios of data center interconnections only up to a few hundred meters, Ozolins and coworkers [10] show up to 100 GBaud capacity over 400 m using OOK and PAM-4 modulation and a monolithically integrated, externally modulated laser with 100 GHz 3 dB bandwidth. Equalizer complexity is also taken into account, studying its evolution as a function of the baud rate. They also show that around 1 km dispersion uncompensated 100 GBaud transmission can be achieved with a simple decision feedback equalizer with a 1 dB received power penalty with respect to back-to-back configuration.

Acknowledgments: The editor would like to thank all the authors for their valuable contribution to the scientific community, all the peer reviewers for taking the time to improve the quality of the submitted manuscripts and, by extension, of the whole Special Issue and editorial team of *Applied Sciences* for their hard work. My gratitude goes especially to Section Managing Editor Chloe Liu for the dedication and continued support shown.

Conflicts of Interest: The authors declare no conflict of interest.

References

1. Winzer, P.J.; Neilson, D.T.; Chraplyvy, A.R. Fiber-optic transmission and networking: The previous 20 and the next 20 years [Invited]. *Opt. Express* **2018**, *26*, 24190–24239. [CrossRef] [PubMed]
2. Morsy-Osman, M.; Sowailem, M.; El-Fiky, E.; Goodwill, T.; Hoang, T.; Lessard, S.; Plant, D.V. DSP-free 'coherent-lite' transceiver for next generation single wavelength optical intra-datacenter interconnects. *Opt. Express* **2018**, *26*, 8890–8903. [CrossRef]
3. Zhong, K.; Zhou, X.; Huo, J.; Yu, C.; Lu, C.; Lau, P.T.A. Digital Signal Processing for Short-Reach Optical Communications: A Review of Current Technologies and Future Trends. *J. Light. Technol.* **2018**, *36*, 377–400. [CrossRef]
4. Plant, D.V. Self-homodyne system for next generation intra-datacenter optical interconnects. *Opt. Express* **2017**, *25*, 27834–27844. [CrossRef]
5. Li, Z.; Zeng, Y.; Song, Y.; Zhang, J.; Zhou, Y.; Ning, Y.; Qin, L.; Wang, L. Effect of Substrate Misorientation on the Structural and Optical Characteristics of In-Rich InGaAs/GaAsP Quantum Wells. *Appl. Sci.* **2021**, *11*, 8639. [CrossRef]
6. Zah, C.-E.; Bhat, R.; Pathak, B.; Favire, F.; Lin, W.; Wang, M.; Andreadakis, N.; Hwang, D.; Koza, M.; Lee, T.-P.; et al. High-performance uncooled 1.3-μm Al/sub x/Ga/sub y/In/sub 1-x-y/As/InP strained-layer quantum-well lasers for subscriber loop applications. *IEEE J. Quantum Electron.* **1994**, *30*, 511–523. [CrossRef]
7. de Farias, A.L.S.; Almeida, R.C.; Chaves, D.A.R. Traffic-Based Heuristics for Regenerator Placement in Translucent Dynamic Optical Networks. *Appl. Sci.* **2022**, *12*, 4388. [CrossRef]
8. Hadi, M.U.; Awais, M.; Raza, M.; Ashraf, M.I.; Song, J. Experimental Demonstration and Performance Enhancement of 5G NR Multiband Radio over Fiber System Using Optimized Digital Predistortion. *Appl. Sci.* **2021**, *11*, 11624. [CrossRef]
9. Wang, H.; Torres-Ferrera, P.; Rizzelli, G.; Mercinelli, R.; Ferrero, V.; Gaudino, R. 200 Gbps/λ PON Downstream C-Band Direct-Detection Links with ≥29 dB Power Budget. *Appl. Sci.* **2022**, *12*, 3538. [CrossRef]
10. Ozolins, O.; Pang, X.; Udalcovs, A.; Schatz, R.; Spolitis, S.; Bobrovs, V.; Jacobsen, G.; Popov, S. 100 Gbaud On–Off Keying/Pulse Amplitude Modulation Links in C-Band for Short-Reach Optical Interconnects. *Appl. Sci.* **2021**, *11*, 4284. [CrossRef]

applied sciences

MDPI

Article

Traffic-Based Heuristics for Regenerator Placement in Translucent Dynamic Optical Networks

Andre L. S. de Farias [1], Raul C. Almeida, Jr. [2] and Daniel A. R. Chaves [1,*]

1 Computer Engineering Department, Polytechnic School of Pernambuco, University of Pernambuco, Recife 50740-540, PE, Brazil; farias.andre@outlook.com
2 Department of Electronics and Systems, Federal University of Pernambuco, Recife 50740-540, PE, Brazil; raul.almeidajunior@ufpe.br
* Correspondence: darc@ecomp.poli.br

Abstract: We propose in this paper two heuristic strategies to solve the regenerator placement problem (RPP) in translucent networks under dynamic traffic. The heuristics are based on both the forecast of the offered load and estimates of blocking probabilities in the network nodes. The proposed heuristics are compared to other regenerator placement algorithms from the literature in two different topologies. The results show that one of the proposed algorithms outperforms, in the investigated scenarios, all other algorithms from the literature considered for comparison purposes in this paper, whereas the second proposed algorithm outperforms the algorithms from the literature only in some considered scenarios.

Keywords: translucent optical networks; regenerator placement; dynamic traffic; heuristics; network design; sparse regeneration; 3R; regenerator

Citation: de Farias, A.L.S.; Almeida, R.C., Jr.; Chaves, D.A.R. Traffic-Based Heuristics for Regenerator Placement in Translucent Dynamic Optical Networks. *Appl. Sci.* **2022**, *12*, 4388. https://doi.org/10.3390/app12094388

Academic Editor: Giuseppe Rizzelli

Received: 27 February 2022
Accepted: 21 April 2022
Published: 27 April 2022

1. Introduction

The technology of dense wavelength division multiplexing (DWDM) divides the available fiber spectrum into wavelength bands named as channels [1,2]. The optical networks that use DWDM technology and perform optical-electrical-optical (O/E/O) conversion in all intermediate nodes of each lightpath are known as opaque networks [1]. These networks usually show a good performance in terms of blocking probability (BP) but require high capital expenditure (CapEx) for their installation due to the large amount of transponders deployed in their nodes [3,4]. In transparent all-optical networks, on the other hand, the O/E/O is not performed in the lightpaths' intermediate nodes, which results in a lower implementation CapEx. However, the signal may be excessively degraded due to physical-layer penalties. These penalties may be mitigated by the use of 3R regenerators [1]. Such regenerators are used whenever the accumulated penalties make the signal quality of transmission (QoT) unacceptable [1,4]. The QoT is usually verified by using either the optical signal to noise ratio (OSNR) [4] or the maximum optical reach (MOR) [2]. MOR is the maximum distance that the optical signal can propagate, keeping the bit error rate (BER) under a certain limit.

The use of 3R alongside a lightpath divides it into transparent segments (TS). The signal propagates with no O/E/O conversions along each TS. The networks that use 3Rs in some of their routes are known as translucent optical networks [1]. These networks aim to find a good CapEx/performance trade-off, drawing the main advantages of both opaque and transparent networks. The algorithm responsible for dividing the lightpath into TS segments is known in the literature as the regenerator assignment (RA) procedure [5–7].

The design of translucent optical networks is a multi-objective problem that needs to deal with two conflicting aspects: the capital cost and the performance improvement. The performance improvement generated by the inclusion of 3R regenerators in the network can be evaluated by reducing the network blocking probability. The cost aspect can be evaluated by the number of regeneration devices installed in the network. The capital cost related to including a large number of regenerators can be of the same order of magnitude as the capital cost required to install all other equipment needed to build the entire optical network [8]. For this reason, the network carriers tend to use as few regenerators as they can in their networks to reduce their costs.

To perform the design of translucent networks, it is important to consider the type of traffic embedded in the network. One can broadly classify the types of traffic as deterministic (also known as static) and stochastic (also known as ad-hoc or dynamic) [9]:

- deterministic/static traffic—the parameters of all network demands are known in advance and they are deterministic information. These parameters for each demand are the source and destination nodes, setup time (start time of the connection), and the disconnection time (end time of the connection). The list of all demands with their information is the traffic matrix.
- ad-hoc/dynamic traffic—the parameters of all network demands, such as source and destination nodes, setup time, and disconnection time, are all unknown, i.e., they are random variables. The clients' connections are requested to the network, on-demand, in an intempestive manner. In this case, usually, the traffic matrix consists of the list of values for each source and destination node used to state stochastic modeling for the traffic.

The design of a translucent network often uses one of two approaches: islands of transparency or sparse distribution [1]. In using the sparse distribution, the network design determines how the regenerators should be distributed among the network nodes. This problem is known in the literature as the regenerator placement problem (RPP). There are three main approaches to solving the RPP [9–11]:

- RPP-1—given a physical topology and a deterministic traffic matrix (static traffic), find the minimum number of regenerators that should be placed in the network nodes to embed in the network all demands present in the traffic matrix, such that all demands can be routed in the network using a path formed by TSs smaller than MOR.
- RPP-2—Given a physical topology and source-destination pairs of all demands listed in the traffic matrix, find the minimum set of network nodes that must have regenerators such that all demands may be routed in the network using a path formed by TSs smaller than MOR.
- RPP-3—Given a physical topology, an ad-hoc traffic matrix and a total number of regenerators that should be deployed in the network, find the number of regenerators that should be placed in each network node to minimize the network blocking probability. In RPP-3, the regenerators are deployed in each network node and they are assigned (used) to each path, on-demand, according to the necessity of the incoming lightpath to be divided into TSs smaller than MOR. Thus, in a given time, there are both assigned (used) and unassigned (free) regenerators in each network node. Eventually, a demand that requires a route longer than the MOR (i.e., it requires regeneration) may be routed throughout nodes with no free regenerators, which causes blocking of the demand.

The three mentioned RPP versions are computationally complex and classified either as NP-complete or NP-hard [4,9–11]. Thus, heuristics [3,12] or meta-heuristics [4,11,13] are often used to solve RPP. Note, however, that the minimization target of RPP-1 and RPP-2 has a linear characteristic (either sum of the number of regenerators deployed or sum of nodes with regeneration capability) and, for this reason, there are several approaches in the literature using (M)ILP to solve either RPP-1 [10,14] or RPP-2 optimally [9,15–20]. Typically, however, (M)ILP methods solve the problem for networks with a low number of nodes [9,10,15–17]. In RPP-3, on the other hand, the minimization target is the network blocking probability. Clearly, the network blocking probability has a non-linear dependence on the number of regenerators deployed in each network node which prevents the use of (M)ILP methods to solve the RPP-3 problem optimally. To optimally solve the RPP-3,

BP modeling is required. Very few BP modelings for translucent networks are available in the literature and they assume several simplifications to reach a very complex (and computationally intensive) set of formulas [21]. This fact prevents using an optimization method over this set of equations to solve RPP-3 in an acceptable time optimally.

In this paper, we are interested in addressing networks under ad-hoc traffic. Clearly, RPP-1 can only solve the design of networks under static traffic. RPP-2, on the other hand, can be used to resolve networks with ad-hoc traffic. However, it cannot do the optimization to reduce BP, nor with the objective of reducing the network implementation cost because it only defines which nodes in the network must have regenerators without answering how many regenerators should actually be installed on each node. Thus, the only approach capable of handling the optimization of both performance (BP) and capex (cost) for networks under ad-hoc traffic is the RPP-3. In this paper, we focus on solving RPP-3, hereafter named as RPP, under ad-hoc traffic. For this reason, we used, in this paper, the heuristic approach to solve the RPP.

The Poison/Markovian process using exponential distribution for connection request inter-arrival and holding times are largely assumed in the literature to model the ad-hoc/dynamic traffic in optical circuit-switched networks [1,2,11–13,16,21–24]. Nevertheless, some classical works [25] bring to attention the self-similar characteristics of data networks traffic, pointing out the inadequacy of the Poison/Markovian process to characterize such network traffic. However, based on observation of actual data traffic, more recent papers have shown that the Poison/Markovian process can closely represent the traffic of current high-speed backbone networks data traffic, mainly because the burstiness of the inter-arrival times decreases significantly in current backbone networks [26–28].

There are two main approaches to solving RPP-3 under ad-hoc traffic heuristically: (1) using either topological or traffic characteristics of the network (such as the degrees of the nodes, number of routes that pass through a node, etc.) to infer the amount of regenerators that should be placed in each node to reduce the BP or (2) using the BP returned by a network simulator and applying this information as a guide during the regenerator placement. The disadvantage of the strategies that use the first approach [1,2,4,12,22,29] is that they do not consider BP information during the RPP solution.

Strategies that use the second approach [11,13,30], on the other hand, take into account BP information during the RPP decision. However, they require a network simulator to return those BPs, which may lead to large complexity and excessive time to reach the RPP solution. In this paper, we propose solving the RPP using the second approach but dismissing the use of network simulations to return the related BPs during the RPP procedure. Instead, for the BP evaluation, we use approximate analytical modeling, which allows the evaluation of the impact of the inclusion of each regenerator in the overall network BP.

Using the heuristic approach, Pedro [2] proposes a statistical framework to tackle RPP in translucent optical networks under ad-hoc traffic. The framework consists of two parts: first, some heuristic is used to evaluate the probability $pdf_{3R}(i)$ of a node i being used as a regeneration point by some lightpath. Then, the regenerators are iteratively placed in the network, one per iteration, in the node that presents the maximum value of the following deviation: $\Delta pdf_{3R}(i) = pdf_{3R}(i) - n_{3R}(i)/n_{3R}^{total}$, in which, $n_{3R}(i)$ is the current number of regenerators deployed in node i and n_{3R}^{total} is the total number of regenerators to be deployed in the network.

In this paper, we propose two heuristics to solve RPP. We introduce two alternative approaches based on Markov modeling to evaluate $pdf_{3R}(i)$, and we propose that during the iterative process, the decision to place a regenerator on a node be based on the BP reduction generated by that choice. To infer the BP reduction, we used a predicted ad-hoc traffic matrix used to model the stochastic behavior of the ad-hoc traffic. Different from the algorithms presented in [2], the proposed strategies can also consider non-uniform traffic. We also introduce the concept of the essential node, which helps solve RPP. Thus, the main contribution of our method is that we approximately evaluate the reduction of the overall network blocking probability (BP) caused by the addition of a regenerator in each network node and we select, for the 3R inclusion, the node that provides the greatest reduction in BP. The previous methods, on the other hand, do not perform such evaluation: they usually place regenerators proportionally to the traffic offered to each node, which does not correspond to the solution that results in the greatest reduction of the BP because the relationship between the reduction in BP and node offered load is not linear.

The paper is organized as follows: in Section 3, we establish some definitions and mathematical notation used in the paper, in Section 4, we introduce the proposed RPP heuristics, in Section 5, we state the simulation setup used in the simulations, in Section 6 we show the simulation results and in Section 7, we give our conclusions.

2. Related Works

In this work, we propose two new approaches to perform sparse regenerator placement in translucent optical networks under ad-hoc/dynamic traffic. Thus, we deal with the problem defined in Section 1 as RPP-3. Note, however, that there are few works in the literature that fully realize the RPP-3 goals by using the network blocking probability information to guide the regenerator placement itself. In this section, we review some algorithms proposed in the literature seeking the solution to RPP-3. Two main types of algorithms are reviewed, the algorithms that fully accomplish the RPP-3 goals (RPP-3d) as well as those that do not use the blocking probability information directly during the placement decision but, instead, apply heuristics whose ultimate goal is to reduce the network blocking probability (RPP-3h).

Yang and Ramamurthy [1] propose four different heuristic algorithms to solve the RPP. Each heuristic assigns a score to each network node and the N best-scored nodes receive a predetermined number of X regenerators (N-X policy). The proposed NDF and CNF algorithms give the score for each network node based, respectively, on the node degree and on how centered in the network the node is. These are topological-based heuristics because only information from the network topology is used during the placement decision. The work also proposed the TLP and SQP traffic-based heuristics. They are traffic-based heuristics because a set of dynamic call requests are simulated to evaluate each network node score. The TLP scores the network node i based on the number of times that a call request is routed through the node i and the SQP does the same but considering the transparent reach in terms of the number of hops.

Some other traffic-based placement proposals use a similar idea and a network simulation engine to simulate a set of dynamic call requests. The statistics of node usage during the simulations are used to score the network nodes towards a regenerator placement decision. Applying this idea, Chaves et al. [4] proposed the MU and MSU heuristics and Walkowiak et al. [29] proposed the SAUR heuristic. The MU scores each node i based on the number of times that the node i is assigned as a regeneration point during the simulation, whereas the MSU algorithm scores each node i based on the maximum number of regenerators simultaneously assigned in the node i during the simulation. In both cases, the total number of regenerators to be deployed in the network are distributed among the network nodes proportionally to each node score. The SAUR heuristic further improves the MSU idea by taking into account not only the number of simultaneously used regenerators in a node but also the statistics regarding the usage.

As mentioned in Section 1, Pedro [2] proposes a statistical framework to tackle RPP-3h in translucent optical networks under ad-hoc traffic. The mentioned $pdf_{3R}(i)$ function is evaluated accordingly to the selected heuristics. He proposes to evaluate the $pdf_{3R}(i)$ considering only topological information such as node degrees and shortest paths between each pair of nodes. Thus, only the topological-based placement is considered in the work. The $pdf_{3R}(i)$ is evaluated using four different strategies: (1) equal value for all nodes (UN); (2) proportionally to the node degree (ND); (3) proportionally to the number of shortest paths that pass through the node i (RO); (4) the same as RO but also considering the transmission reach (RR).

Moreover, applying topological-based placement, Aibin and Walkowiak [12] propose the DA heuristic. This heuristic also assigns a score to each network node. The score of the node i is evaluated by the quotient between the sum of the lengths of the links connected to the node i by the total length of all links in the network. Then, the regenerators are distributed proportionally to each node score.

The algorithms reviewed so far apply the RPP-3h approach. On the other hand, Cavalcante et al. [11] and Chaves et al. [13] apply a genetic-algorithm-based solution to solve a multi-objective version of the RPP-3d problem for, respectively, elastic and WDM optical networks. Based on the forecast traffic matrix, the algorithm seeks to find the regenerator placement solution which achieves the best trade-off between the total number of deployed regenerators and the network blocking probability. These algorithms reach lower blocking probabilities than other placement algorithms in the literature. The traffic information and the total network blocking probability are used to influence the placement decision itself, but a network simulator engine is needed to assess each solution proposed by the genetic algorithm during the regenerator placement procedure.

Zhao et al. [21] propose analytical models for computing the blocking performance of translucent optical networks with sparsely located regeneration nodes. The analytical models developed to estimate the call blocking probability in an optical network are applied to assess two regeneration node allocation policies. The proposed analytical models are suitable to be used to assess the translucent optical network performance, but they are computationally complex enough to prevent its use during the translucent network design (regenerator placement decision) because several different placement combinations must be investigated during the designing procedure.

The results from the above-reviewed papers consistently show that traffic-based placement approaches achieve lower blocking probabilities in comparison to topology-based approaches. Traffic-based approaches can also satisfactorily solve the placement problem for the network under both uniform and non-uniform traffic patterns, whereas the topology-based approaches give the same result regardless of the offered traffic pattern. All reviewed traffic-based approaches require a network simulator engine to simulate a realistic per node regenerator demand (in the case of the application of RPP-3h) [1,4,29] and its resultant network blocking probability (in the case of the application of RPP-3d) [11,13]. Different from the other algorithms in the literature, we propose in this paper an approximate analytical model to evaluate the network blocking probability and we employ this model during the placement procedure. Our heuristic proposals use the blocking probability information to guide the regenerator placement. The proposed heuristics are traffic-based placement algorithms that can, at the same time, heuristically solve RPP-3d, deals with uniform and non-uniform traffic patterns, and dismiss the use of a simulator engine during the regenerator placement decision.

3. Definitions

This section specifies the notations used to describe the proposed RP heuristics presented in Section 4.

3.1. Premises and Notation

Suppose a network with N nodes and MOR given by T_R. The route between nodes i and j is defined as $\pi_{ij} = [n_1, n_2, \ldots, n_{H_{ij}}]$, where H_{ij} is the number of nodes in the route. Let us assume that π_{ij} has a total physical length T_{ij} and a total optical length $V_{ij} = T_{ij} + (H_{ij} - 2) \cdot \epsilon$. The optical length accounts for the physical length added by an extra ϵ length per intermediate node, representing the degradation suffered by the signal passing through these nodes [2]. Using π_{ij}, V_{ij} and T_R, it is possible to determine the minimum number of regenerators, K, that keeps the optical length of all $K + 1$ resultant TSs in π_{ij} shorter than T_R.

For a network under ad-hoc traffic, lightpaths are set up and torn down dynamically. In these networks, the traffic is usually assumed memoryless and evaluated in an erlang unit, defined as the ratio between the average connection holding time and the average time between requests. The ad-hoc traffic matrix is represented in this paper as $\Omega = \{\omega_{ij}\}$, such that ω_{ij} represents the traffic between nodes i and j (in erlang unit), with $\omega_{ij} = 0$ if $i = j$ and the total network load is $L = \sum_{i=1}^{N} \sum_{j=1}^{N} \omega_{ij}$. The traffic matrix that considers only those paths π_{ij} that require mandatory regeneration (i.e $V_{ij} > T_R$) is $\Gamma = \{\gamma_{ij}\}$, in which $\gamma_{ij} = \omega_{ij}$ if $V_{ij} > T_R$ and $\gamma_{ij} = 0$ otherwise.

3.2. Set of Possible Regenerator Combinations (PRC)

A regenerator combination of a route is a combination of nodes in the route that satisfies the criteria that all TSs formed with the signal regeneration in such nodes are feasible, i.e., their optical lengths are shorter than T_R. Let C_{ij} be the set of all those regenerator combinations on route $\pi_{i,j}$ that use the minimum number of possible regeneration points, K_{ij}, i.e.,:

$$\begin{aligned} C_{ij} =&[(n_1^{(1)},\ldots,n_{K_{ij}}^{(1)})\ldots(n_1^{(M_{i,j})},\ldots,n_{K_{ij}}^{(M_{i,j})})] \\ =&\{c_{ij}^{(1)},\ldots,c_{ij}^{(M_{i,j})}\}, \end{aligned} \tag{1}$$

in which $M_{i,j}$ is the number of regenerator combinations of $\pi_{i,j}$ and $n_k^{(m)}$ is the k-th selected regeneration node along the route π_{ij} of the m-th regenerator combination. C_{ij} is defined as the possible regenerator combinations (PRC). It stores all regenerator combinations that segment the route π_{ij} into $K_{ij} + 1$ feasible transparent segments. Notice that $M_{ij} \leq \binom{H_{ij}-2}{K_{ij}}$, since $H_{ij} - 2$ is the number of intermediate nodes where regeneration can take place and K_{ij} is the number of nodes using regeneration.

3.3. Set of Feasible Regenerator Combinations (FRC)

In a typical translucent network, some nodes in the network may not have installed regenerators. In addition, even in nodes with installed regenerators, they may all be busy (i.e., in use) on the arrival of a connection request. Thus, a possible regenerator combination (as defined in Section 3.2) is referred to as feasible only if all of its nodes have at least one regenerator installed and it is free on the arrival of a request. We may define a subset $D_{ij} \subseteq C_{ij}$ that considers only the feasible regenerator combinations in C_{ij}. D_{ij} is formed by $M'_{ij} \leq M_{ij}$ feasible regenerator combinations in C_{ij}, described as

$$\begin{aligned} D_{ij} =&[(n_1^{(1)},\ldots,n_{K_{ij}}^{(1)})\ldots(n_1^{(M'_{ij})},\ldots,n_{K_{ij}}^{(M'_{ij})})] \\ =&\{d_{ij}^{(1)},\ldots d_{ij}^{(M'_{ij})}\}, \end{aligned} \tag{2}$$

in which $d_{ij}^{(m)}$ is one of the feasible regenerator combinations. If H'_{ij} is the number of intermediate nodes in π_{ij} with at least one installed and free regenerator (note that $H'_{ij} \leq H_{ij} - 2$), we have that $M'_{ij} \leq \binom{H'_{ij}}{K_{ij}}$.

3.4. Essential Nodes

Using the PRC set, it is possible to define the concept of essential node. Node n_u is defined as essential for the route π_{ij} if $n_u \in c_{ij}^{(m)}$ for all values of m ($1 \leq m \leq M_{ij}$). In other words, n_u is essential if it participates in all PRCs of π_{ij}. Note that it is mandatory to place regenerators in essential nodes because, otherwise, some routes become unfeasible in terms of QoT. For this reason, the proposed algorithms in this work start by placing regenerators in the essential nodes.

4. Proposed Regenerator Placement Heuristics

The regenerator placement heuristics proposed in this paper are based on: (1) defining either the possible or feasible regenerator sets for each source-destination node pair, as described in Sections 3.2 and 3.3, respectively; (2) estimating the load offered to each node by considering only the load of the routes which demand regeneration, and (3) Given the estimated load per node and the number of regenerators installed per node, estimate the overall network blocking probability mitigation when a new regenerator is installed in a specific node.

In this section, we propose two strategies to solve RPP. The proposed strategies coincide in steps (1) and (3) and differ in the form of how the offered load to each network node is estimated. We describe in Section 4.1 how the offered load to a node is estimated in the proposed strategies. It is important to mention that the load and blocking probabilities estimates presented in Sections 4.1 and 4.2 are not intended to give an exact evaluation of such values. Instead, they are intended to be used as an approximate, although precisely enough, estimate so that an efficient heuristic approach to solve RPP may be built.

Although both proposals consider the network traffic pattern forecast and intensity as input information to solve RPP; they can provide efficient regenerator placement even when the actual network traffic is moderately different from that one forecast, as discussed in the results section.

4.1. Estimation of the Total Regeneration Load Offered to a Node

The load offered to the pool of regenerators installed in the network node n can be roughly estimated by the sum of the loads of all lightpaths that need regeneration and pass through that node. This sum can be evaluated as

$$\beta_n = \sum_{i=1}^{N}\sum_{j=1}^{N} G_{ij}(n), \tag{3}$$

in which $G_{ij}(n) = \gamma_{ij}$ if $\exists\, m \,/\, n \in c_{ij}^{(m)}$ and 0 otherwise.

It is possible to improve the β_n estimate by considering that the offered load γ_{ij} is proportionally distributed among all FRCs. The reasoning for this is that two FRCs cannot be used simultaneously. For instance, if there are three feasible combinations for π_{ij}, then the offered load for each combination is approximately $\gamma_{ij}/3$. The estimate of the total effective offered load to node n can be evaluated by

$$\alpha_n = \sum_{i=1}^{N}\sum_{j=1}^{N} F_{ij}(n), \tag{4}$$

in which

$$F_{ij}(n) = \begin{cases} \frac{\gamma_{ij}}{M'_{ij}} \sum_{m=1}^{M'_{ij}} P_{ijm}(n), & \text{if } n \in \pi_{ij} \\ 0, & \text{otherwise,} \end{cases} \tag{5}$$

with $P_{ijm}(n) = 1$ if $n \in d_{ij}^{(m)}$ and 0 otherwise. In this paper, it is proposed to evaluate the $pdf_{3R}(n)$ (using the notation from [2]) by either: $\beta_n / \sum_{i=1}^{N} \beta_i$ or $\alpha_n / \sum_{i=1}^{N} \alpha_i$.

4.2. Blocking Probability Estimation

Consider a node n with $r(n)$ regenerators installed and the estimate of its offered load given by Γ_n (Γ_n as a place holder for either α_n or β_n). Node n can be seen as a pool of $r(n)$ servers under an offered load Γ_n. In such a case, the probability that node n has no available regenerator on the arrival of a request can be evaluated as: $BP(n) = B(r(n), \Gamma_n)$, in which $B(s, a)$ is the Erlang B formula, derived using the Markov chain theory [31], that returns the blocking probability (BP) of a system with s servers under an offered load a.

To choose which node the regenerator shall be installed on, it is important to quantify the impact that the addition of a regenerator in each network node causes on the network path request blocking probability. As known in Markov theory, the blocking probability of a node n, $BP(n)$, is the long-term proportion of time that the node spends without an available regenerator. Therefore, the number of requests that require a regenerator at node n and are blocked is proportional to $BP(n)$ multiplied by the total requested traffic. We may then think about quantifying the portion of the overall network BP that is caused by blocking imposed on the paths that use node n as a regeneration point.

Let us assume that each path π_{ij} that may use a regenerator in node n experiences approximately a blocking probability given by $BP(n)$. Using $E_{ij}(n)$ as a place holder for either $G_{ij}(n)$ or $F_{ij}(n)$, it is possible to estimate the amount of the overall network BP only due to the blocking caused by node n as

$$BP_N(n) = \frac{\sum_{i=1}^{N}\sum_{j=1}^{N} B(r(n),\Gamma_n)\cdot E_{ij}(n)}{\sum_{i=1}^{N}\sum_{j=1}^{N}\omega_{ij}} \quad (6)$$

or

$$BP_N(n) = (\Gamma_n/L)\cdot B(r(n),\Gamma_n). \quad (7)$$

Thus, one may evaluate the network BP reduction caused by the inclusion of a single regenerator in the node n using

$$\Delta BP_N(n) = \frac{\Gamma_n}{L}([B(r(n),\Gamma_n) - B(r(n)+1,\Gamma_n)]). \quad (8)$$

The deviation shown in Equation (8) is used to decide which node receives a new regenerator at each algorithm iteration.

We provide in Appendix A a proposal to evaluate (8) considering a flexgrid/multiple rates demands scenario that occurs in elastic optical networks (EONs).

4.3. *Regenerator Placement Algorithms*

The regenerator placement proposed in this paper is shown in Algorithm 1. The algorithm requires as input parameters the total amount of regenerators R to be deployed in the network, the forecast traffic matrices $\{\omega_{ij}\}$ and $\{\gamma_{ij}\}$ and the network topology. In the algorithm, Γ_n stands for either α_n or β_n and $\Delta(n) = \Delta BP_N(n)$. First, the algorithm identifies and places one regenerator in each essential node. After that, it places one regenerator per iteration. The estimates of node offered load are re-evaluated in each iteration. The node that returns the highest BP reduction (i.e., the highest value for $\Delta(n)$) receives one regenerator. Ties are broken, firstly by placing the regenerator in the node with the largest Γ_n and then randomly. The process is repeated until all regenerators are placed. Regarding the selection of Γ_n, two heuristics are proposed: Fixed Load Distribution (FLD), which uses $\Gamma_n = \beta_n$ and Dynamic Load Distribution (DLD), which uses $\Gamma_n = \alpha_n$. Note that, in the particular case in which $\Delta(n) = \Delta pdf_{3R}(n)$ and under uniform traffic, the FLD is equal to the Routing-&-Reach algorithm [2].

Our algorithm can be classified as traffic-based because the traffic matrices w_{ij} and γ_{ij} directly affect the decision of placing a regenerator in a given node. Please note in Algorithm 1 that the decision of placing a regenerator in a node depends on $\Delta(n)$, which, in the proposed algorithms, is evaluated as $\Delta BP_N(n)$ by using Equations (6)–(8). Note that, in such equations, the estimated offered load Γ_n to a node and the current number of installed regenerators $r(n)$ in a node are taken into account. Finally, w_{ij} directly appears in Equations (6) and (7), and the evaluation of the estimated offered load to a node, Γ_n, takes into account γ_{ij} by using either Equations (3) or (4).

4.4. *Time Complexity Analysis of the Proposed Algorithms*

Let us start the analysis of the time complexity of the proposed algorithms by discussing the time complexity required to evaluate both the M_{ij} and M'_{ij} combinations. As shown in Sections 3.2 and 3.3 for a route π_{ij}, $M_{ij} = \binom{H_{ij}-2}{K_{ij}}$ and $M'_{ij} = \binom{H'_{ij}}{K_{ij}}$. Note that, for each route, the value of K_{ij} is constant and known in advance. In such case, the evaluation of all combinations can be done in polynomial time if K_{ij} is not close to $(H_{ij} - 2)/2$, considering M_{ij} evaluation, and $(H'_{ij}/2)$, considering M'_{ij} evaluation. In a typical mesh optical network, it is expected that the signal may propagate for k hops before a regeneration is required (i.e., usually k is larger than 2) . Therefore, K_{ij} is usually smaller than $H_{ij}/2$, which means that both M_{ij} and M'_{ij} can be evaluated in polynomial time. Even in the case in which K_{ij} is close to either $(H_{ij} - 2)/2$ or $H'_{ij}/2$, the required time to evaluate both the M_{ij} and M'_{ij} combinations is low if H_{ij} is small. All routes with few (i.e., 2 to 5) hops fit in this case regardless of the number of hops the signal may traverse without regeneration. Typically, such cases represent the majority of the routes in a mesh optical network. Moreover, it is expected that the number of hops in a route can be considerably smaller than the number of nodes in the network. This is expected even for long routes. Thus, it could be concluded that the evaluation of the combinations is not prohibitive even for networks with a high number of nodes.

The most time-consuming step in Algorithm 1 is the Γ_n evaluation. It requires, in the worst case, finding the shortest path (SP) between all node pairs ($\mathcal{O}(N^3)$) and, for each SP, the evaluation of either all PRCs (for FLD) or all FRCs (for DLD). By defining $Z = \max_{i,j} M_{ij}$ (maximum number of PRCs) and $Z' = \max_{i,j} M'_{ij}$ (maximum number of FRCs) we may write that the DLD shows $\mathcal{O}(R \cdot N^3 \cdot Z')$ complexity whereas FLD (and Routing-&-Reach) shows $\mathcal{O}(N^3 \cdot Z)$. Note that DLD requires Γ_n reevaluation after each 3R inclusion and FLD does not. Given the above time complexities and the fact that $M'_{ij} \leq M_{ij}$, we can say, for the sake of comparison, that DLD requires, in the worst case, R times more time to solve the RPP problem than FLD.

Algorithm 1 Pseudocode for proposed heuristics

```
Require: Total amount of regenerators to be deployed R
Require: Traffic matrices forecast { ω_ij } and { γ_ij } and the topology
  r ← R
  Determine all essential nodes
  e ← Amount of essential nodes
  r ← r − e
  Place one regenerator in each essential node
  while r > 0 do
    for n = 1 to N do
      Place a temporary regenerator in node n
      Evaluate Γ_n of all nodes
      Evaluate Δ(n)
      Remove the temporary regenerator placed at node n
    Place one reg. in node n with highest value Δ(n)
    r ← r − 1
```

5. Simulation Setup

We use two topologies to evaluate the performance of the RPP algorithms: CoreNet (75 nodes and 99 bi-directional links) [32] and EuroNet (28 nodes and 41 bidirectional links) [33], both with 96 wavelengths per link, each transmitting 100 Gb/s PDM-QPSK. In the simulations, the transmission rate only affects the MOR and, unless stated otherwise, it is used the same values for 100 Gb/s adopted by Pedro [2]: $T_R = 2000$ km, and $\epsilon = 60$ km. However, we have also investigated the cases for $T_R = 1600$ km, $\epsilon = 60$ km and $T_R = 2400$ km, $\epsilon = 60$ km to validate the effectiveness of the proposed method under different transmission reaches.

Uniform traffic (UT) and non-uniform traffic (nUT) are considered during simulations. Under uniform traffic, $\omega_{ij} = L/(N \cdot (N-1))$ and under non-uniform traffic,

$\omega_{ij} = (\sigma_{ij} \cdot a \cdot L)/(N \cdot (N-1))$. The σ_{ij} is a random number uniformly distributed on the interval $[0.5; 2.5]$ and the constant a is such that: $a = (N \cdot (N-1))/(\sum_{i=1}^{N}\sum_{j=1}^{N}\sigma_{ij})$. The random $\sigma_{ij} \in [0.5; 2.5]$ means that the load between the nodes i and j for nUT traffic may vary from 50% to 250% of the load for the same pair under UT traffic, maintaining; however, the same total network load for both traffics.

There are two approaches mentioned in the literature and used in the industry to perform 3R in optical networks: (1) a shared pool of regenerators per node [1,2,11]; (2) the use of not assigned (i.e., free) add-drop interfaces to perform back-to-back regeneration [23]. In this paper, we assume that there is a pool of shared devices in each node exclusively dedicated to providing 3R capability and they can perform both wavelength conversion and signal regeneration.

The node architecture assumed in this paper is shown in Figure 1. It is composed by a colorless/directionless/contentionless (CDC) ROADM, de/multiplexers, add/drop interfaces, and a shared bank of 3R regenerators. Using this architecture, it is possible to decide, by using the ROADM switch fabric, whether the optical signal should be added/dropped, be bypassed, or undergo to the 3R shared bank to be regenerated. The node shown in Figure 1 supports a pair of fibers in each direction, the fibers arriving at the node are connected to ROADM switching fabric through the DEMUXes, whereas the fibers leaving the node are connected via MUXes. For example, signals from a client/user connected to a network node can be transmitted (or received) by accessing the "Add/Drop interfaces" block (bottom left in the figure). Then, they can be optically switched by ROADM to one of its outputs (or inputs), connected to a MUX (or DEMUX), until the signal may reach one of the output (or input) fibers of the node. On the other hand, an optical signal that is being transmitted along with the network and that arrives at a node (via fiber + DEMUX input) can be switched to an output port (via fiber + MUX output), or it can undergo electronic regeneration, being switched by the ROADM to the shared 3R regenerator bank (bottom right in the figure). In this case, the signal is regenerated by one of the 3R regenerators available in the bank and then inserted again into one of the input ports of the ROADM to be forwarded to one of the output fibers (fiber + MUX).

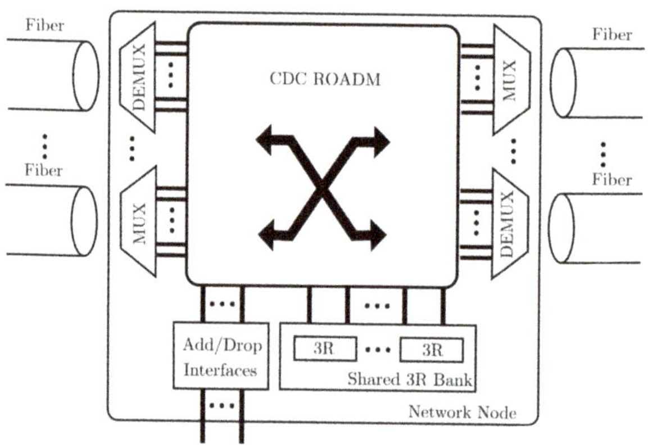

Figure 1. Node architecture assumed in this paper composed by a colorless/directionless/ contentionless (CDC) ROADM, de/multiplexers, add/drop interfaces, and a shared bank of 3R regenerators (adapted from [2]).

The call requests are generated under a Poisson stochastic process in which the call duration follows an exponential distribution. Upon a call request, the shortest path (in terms of distance) from the source to the destination node is found using Dijkstra's algorithm (we also investigated some cases using the k-shortest path routing algorithm as discussed in Section 6). The route is then segmented into TSs using the same RA used in [4] considering the MOR instead of OSNR as the QoT criterion. This RA uses regenerators to make feasible routes with issues in either wavelength continuity constraints or QoT. The first fit algorithm performs wavelength assignment (WA) in each TS. If either RA or WA fails, the call is blocked; otherwise, the call is accepted.

In the blocking probability graphs shown in Section 6, the error bars for a confidence level of 95% are too small to appear in the graphs (smaller than the plot symbols used). It is obtained by repeating the same simulation point 30 times.

6. Results

We analyze in this section the performance of the proposed FLD and DLD algorithms to solve RPP and compare them with the following previously proposed strategies: Uniform (UN), Nodal-Degree (ND), Routing-Only (RO), and Routing-&-Reach (RR) [2]. In brief, all these algorithms distribute the R regenerators to each network node: uniformly (UN), proportional to the node degree of the node (ND), proportional to the number of shortest paths that pass through the node (RO), same as RO but also considering the transmission reach (RR). We also show the BP of the same network topologies, either transparent or opaque. Transparency occurs when no regenerators are installed in the network nodes. Opaqueness is achieved when all nodes are fully equipped with 3Rs. This requires a total of 19,008 regenerators installed in CoreNet and 7812 in EuroNet. In the simulated scenarios considering a transparent network (i.e., zero regenerator installed per node), we verified that the BP is dominated by the blockings due to paths π_{ij} with unacceptable values of T_{ij} (lack of reach), i.e., the blockings due to lack of wavelength are negligible in these scenarios. However, as the number of the deployed regenerators in the network increases, blockings occur due to both lack of reach and lack of wavelength. The relative values between these two blocking causes depend on the number of deployed regenerators.

Figure 2 (Figure 3) shows the BP as a function of the total number of regenerators installed in the CoreNet (EuroNet) topology under a load of 600 erl (500 erl), with Figure 2a (Figure 3a) standing for uniform traffic and Figure 2b (Figure 3b) standing for non-uniform traffic considered during the placement procedure. To evaluate how efficient are the proposed RPP procedures in cases where the traffic pattern used during the RPP process is different from the actual traffic offered to the network, we have investigated two situations: (1) The same traffic pattern is considered in the RPP procedure as well as in the BP evaluation (represented by symbols in the graphs) and (2) the traffic pattern for BP evaluation is different from the one considered during the RPP procedure (lines in the graphs). The case 1 has been named as the EQU scenario and case 2 as the DIF scenario. For each curve in the graphs, the traffic pattern considered during the RPP procedure is indicated after the p = symbol, whereas the traffic pattern considered during the BP evaluation is indicated after the s = symbol. For instance, DLD p = UT s = nUT means that the DLD procedure placed the regenerators considering UT, whereas the BP was estimated for the network under nUT. This is an example of a DIF scenario. Note as well that the UN, ND, RO, and RR algorithms do not take into account any information regarding the traffic matrix, and we have used the notation p = * in such cases.

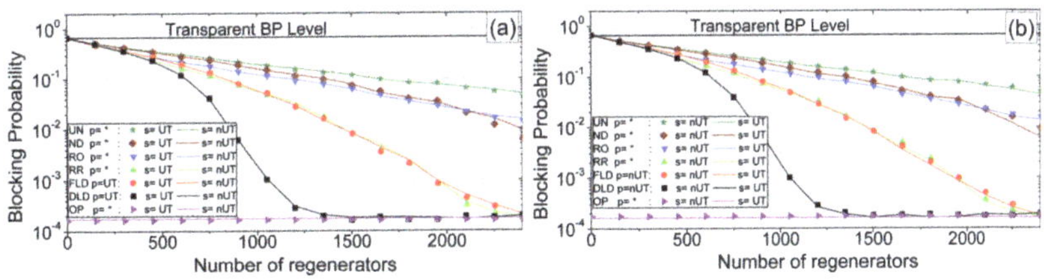

Figure 2. Blocking probability as a function of the total number of regenerators in CoreNet topology under 600 erlangs load and: (**a**) uniform and (**b**) non-uniform traffic.

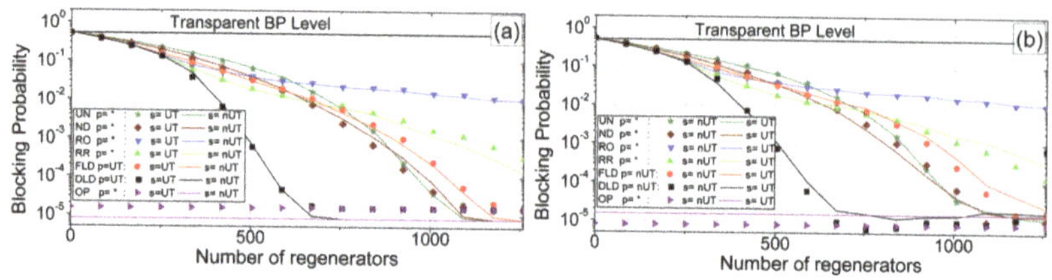

Figure 3. Blocking probability as a function of the total number of regenerators in EuroNet topology under 500 erlangs load: (**a**) uniform and (**b**) non-uniform traffic.

As expected, in Figures 2 and 3, the higher the number of deployed regenerators, the lower the BP found by the algorithms. Eventually, some algorithms reach a saturation point beyond which the placement of more regenerators results in no further reduction in BP. Note in Figures 2 and 3 that the DLD finds solutions that achieve significantly lower values of BP than all other investigated algorithms in both topologies and traffic patterns. Either FLD or RR algorithms find the second lowest values of BP in the entire investigated range in CoreNet and for low to a moderate number of regenerators in EuroNet. For a high number of deployed regenerators in EuroNet, on the other hand, the ND and UN algorithms present lower BP values than both FLD and RR.

Figure 2 (Figure 3) also shows that the DLD strategy can achieve with approximately 1330 (700) regenerators the same performance as an opaque CoreNet (EuroNet) network, whereas the second best algorithm, RR (ND), achieves the same level using 2090 (1100) regenerators. It means that the DLD can reach the opaque network BP level in the CoreNet (EuroNet) topology using only about 7% (9%) of the total regenerators required to mount an opaque network and using 63% (63% for UT and 74% for nUT) of the regenerators required by the second-best algorithm to reach the same level.

Figure 4 (Figure 5) shows, as an example, how the considered RPP algorithms distribute the regenerators along the network nodes in CoreNet (EuroNet) when a total of 1050 (504) regenerators are placed in the network considering UT. The figures show how many regenerators are placed in each node (according to its node ID). In both figures, one can observe that DLD, FLD, RR, and RO algorithms achieve a similar distribution of regenerators along the nodes. However, FLD, RR, and RO tend to promote a high concentration of regenerators in the same specific nodes, whereas DLD tends to promote a slightly more homogeneous distribution of regenerators among the nodes.

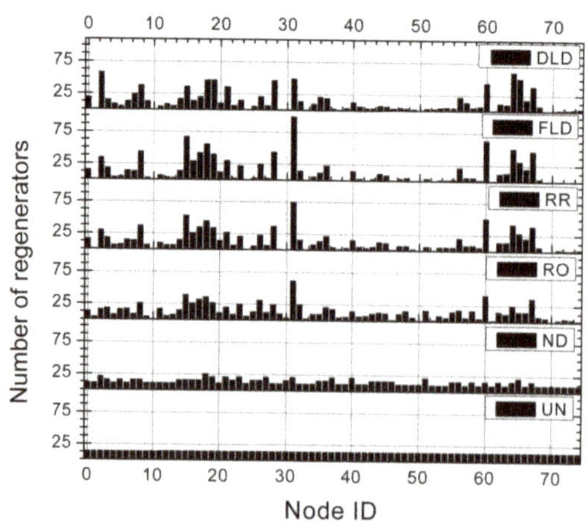

Figure 4. Number of regenerators placed by each considered RP algorithm in each network node of CoreNet assuming the uniform traffic and total amount of 1050 regenerators placed in the network.

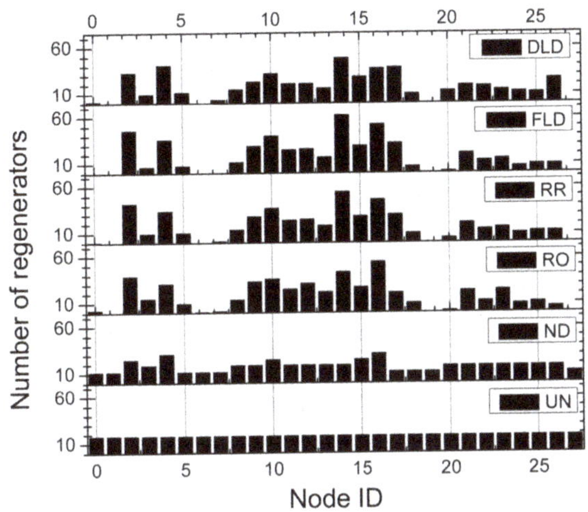

Figure 5. Number of regenerators placed by each considered RP algorithm in each network node of EuroNet assuming the uniform traffic and total amount of 504 regenerators placed in the network.

Figure 6 (Figure 7) shows the BP as a function of the total offered load in the CoreNet (EuroNet) topology under uniform (Figures 6a and 7a) and non-uniform. (Figures 6b and 7b) traffic profiles for the translucent solutions found by each algorithm with a total of 1050 (504) regenerators. Note that the DLD algorithm achieves the lowest BP values among all algorithms for both topologies under either UN or nUT traffic patterns and for the entire investigated load range. Moreover, in both topologies, the relative performance of the algorithms is almost unaltered in terms of BP as the obtained curves show almost no crossing points.

We can also analyze the EQU scenario against the DIF scenario in Figures 2, 3, 6 and 7 by comparing the lines against the symbols of the same color. Note that there are small differences in BP values by making this comparison for a given number of regenerators placed in the network. It means that the RPPs algorithms investigated are very robust against the investigated traffic changes. The premise of our algorithm is to solve the RPP problem to reduce the PB by using the information about the forecast traffic to the network. This is a valid premise as, frequently, there is an approximate information on the network traffic forecast. It makes sense, however, to consider that the actual network traffic be moderately different from the one forecast, but not completely different. With the provided results (i.e., comparison between the DIF and EQU scenarios), we show that, even for moderate differences between forecast and actual network traffic, the proposed RPP algorithms work satisfactorily.

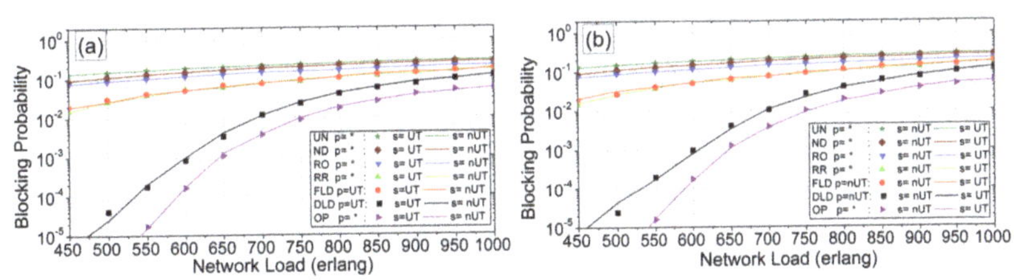

Figure 6. Blocking probability as a function of the total load in CoreNet topology using 1050 regenerators and traffic: (**a**) uniform and (**b**) non uniform.

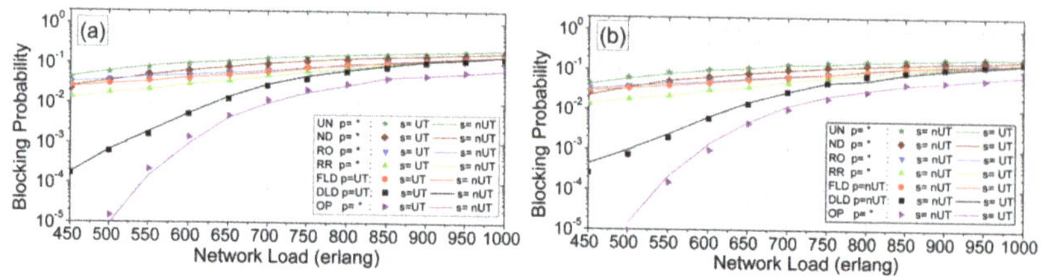

Figure 7. Blocking probability as a function of the total load in EuroNet topology using 504 regenerators and traffic: (**a**) uniform and (**b**) non uniform.

We also investigated the performance of the RPP algorithms when the network operates using the k-shortest path routing algorithm instead of the 1-shortest path (Dijkstra's algorithm) assumed during the RPP procedure. In this case, the simulation is performed by trying, successively, each of the k shortest routes (in terms of distance) found by Yen's k-shortest path routing algorithm. Figure 8 (Figure 9) shows BP as a function of the total number of regenerators installed in the CoreNet (EuroNet) topology. under UT and total offered load of 600 erl (500 erl).

Figure 8a (Figure 9a) stands for the comparison between the BPs verified using $k = 3$ (shown as closed symbols) against the ones verified for $k = 1$ (shown as lines), whereas Figure 8b (Figure 9b) stands for the comparison between the BPs verified using $k = 5$ (closed symbols) against the ones verified for $k = 1$ (lines). As expected, the greater the value of k considered, the lower the BP achieved by a given algorithm. In all investigated cases, the DLD algorithm remains the best cost-effective strategy to solve RPP regardless of the value of k selected. It means that, for a given number o regenerators, the DLD strategy always returns lower or equal BP when compared to the other strategies. Moreover, note that the relative performance of the algorithms, in terms of BP, remains almost unaltered by comparing their BPs results with $k = 1$ against the BPs results using either $k = 3$ or $k = 5$. It means that, for a given value of R, the algorithm that returns the lowest BP for $k = 1$ also returns the lowest BP for both $k = 3$ and $k = 5$. The same occurs with the algorithm that returns the second-lowest BP and so on. There are some exceptions, such as the comparison between ND and RO in CoreNet for a large number of 3R installed.

The performance of the RPP algorithms for different transparent reaches is also investigated. For this purpose, we have decided to show the results for the best (in terms of BP) 4 algorithms investigated so far: ND, RR, FLD, and DLD. We investigate the algorithms performance assuming 3 different transmissions reaches T_R: T_R = 1600 km, 2000 km, and 2400 km. Notice that the T_R is an input parameter for the RPP algorithms RR, FLD, and DLD, as it is taken into account during the RPP procedure. Thus, different values of T_R require different execution of RR, FLD, and DLD.

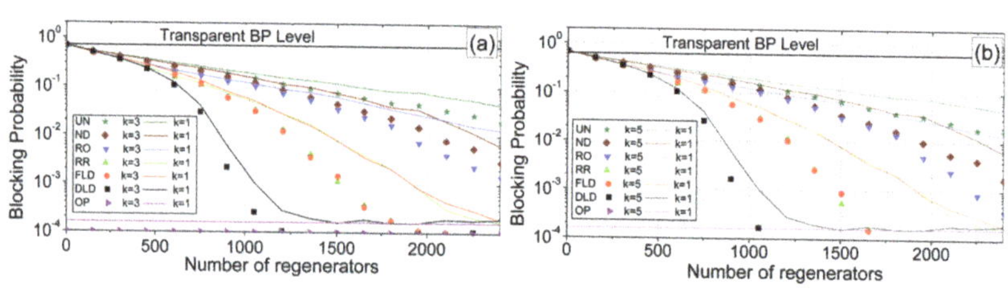

Figure 8. Blocking probability as a function of the total number of regenerators in CoreNet topology using k-shortest path routing algorithm for: (**a**) $k = 3$; (**b**) $k = 5$, each compared against the $k = 1$ case.

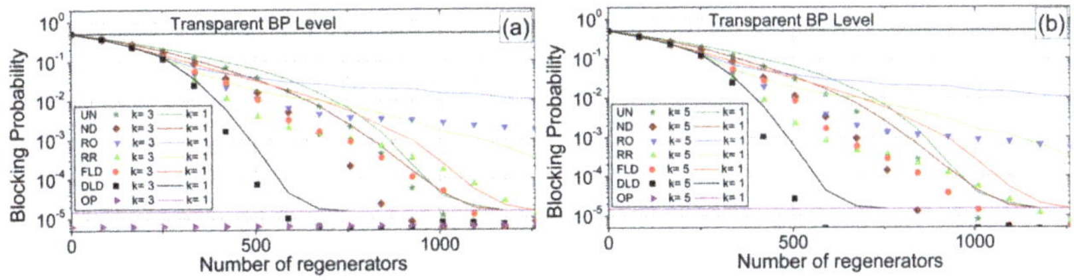

Figure 9. Blocking probability as a function of the total number of regenerators in EuroNet topology using k-shortest path routing algorithm for: (**a**) $k = 3$; (**b**) $k = 5$, each compared against the $k = 1$ case.

Figure 10 (Figure 11) shows BP as a function of the total number of regenerators installed in the CoreNet (EuroNet) topology under a uniform traffic and total offered load of 600 erl (500 erl). Figure 10a (Figure 11a) stands for the comparison between the BP verified assuming T_R = 1600 km (shown as closed symbols) against the one verified for T_R = 2000 km (shown as lines), whereas Figure 10b (Figure 11b) stands for the comparison between the BP verified assuming T_R = 2400 km (closed symbols) against the one verified for T_R = 2000 km (lines).

As expected, the greater the value of T_R considered, the lower the BP achieved by a given algorithm. In all investigated cases, the DLD algorithm remains the best cost-effective strategy to solve RPP regardless of the value of T_R selected. It means that, for a given number o regenerators, the DLD strategy always returns lower or equal BP when compared against the other strategies assuming the same T_R. On the other hand, the relative performance of the algorithms, ND, RR, and FLD, in terms of BP, shows significant dependence on the assumed T_R, mainly for a moderate/high number of regenerators installed in the network. In CoreNet, FLD and RR return almost the same values of BP in the entire considered range for R assuming both T_R = 1600 km and T_R = 2000 km (Figure 10a), whereas RR returns a slightly lower BP than FLD for T_R = 2400 km and moderate values of R (Figure 10b). The most significant inversions are verified for the EuroNet. Assuming T_R = 2000 km, ND returns lower BPs than RR and FLD for moderate to low numbers of R (compare the lines in Figure 11a,b. However, for T_R = 1600 km ND returns the highest value of BP when compared to the ones returned by both RR and FLD for moderate to high values of R. Moreover, whereas FLD returns significantly lower BPs than RR for both T_R = 1600 km and T_R = 2000 km, RR returns slightly lower BP than FLD for T_R = 2400 km.

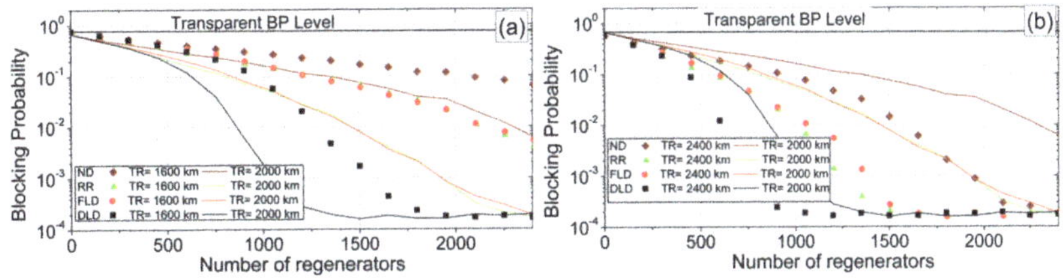

Figure 10. Blocking probability as a function of the total number of regenerators in the CoreNet topology assuming transparent reaches of: (**a**) T_R = 1600 km; (**b**) T_R = 2400 km, each compared against the reach of T_R = 2000 km.

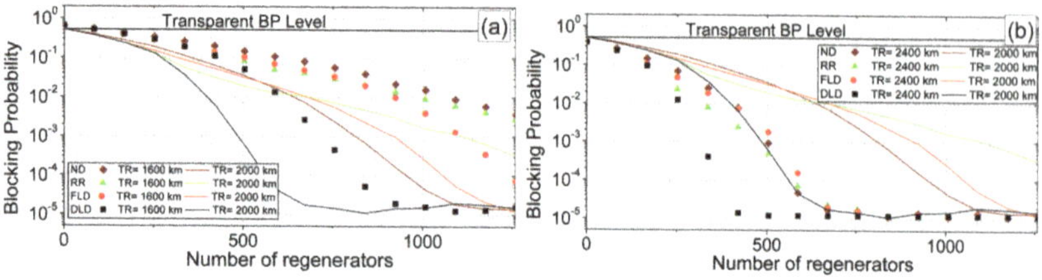

Figure 11. Blocking probability as a function of the total number of regenerators in the EuroNet topology assuming transparent reaches of: (**a**) T_R = 1600 km; (**b**) T_R = 2400 km, each compared against the reach of T_R = 2000 km.

Finally, we also investigated the number and the arrangement of the essential nodes in the analyzed topologies considering the transmission reaches of T_R = 1600 km, T_R = 2000 km, and T_R = 2400 km, as shown in Table 1. In this table, we provide the number of essential nodes in each topology T_R scenario, as well as the node indexes of the essential nodes in the topologies. One can note that, as expected, the number of essential nodes reduces as the transmission reach increases. This occurs because the higher the transmission reach, the lower the number of required regeneration points along a given route.

Table 1. Total amount and node indexes of the essential nodes found for EuroNet and CoreNet topologies considering different transmission reaches of T_R = 1600 km, T_R = 2000 km, and T_R = 2400 km.

Topology	T_R (km)	Number of Essential Nodes	Index of the Essential Nodes
EuroNet	1600	19	2, 3, 4, 5, 6, 7, 9, 10, 11, 12, 14, 16, 17, 19, 20, 21, 23, 24, 26
EuroNet	2000	13	2, 4, 8, 10, 11, 14, 15, 16, 17, 20, 21, 22, 26
EuroNet	2400	4	9, 13, 16, 22
CoreNet	1600	42	0, 2, 3, 6, 7, 8, 11, 12, 13, 14, 15, 16, 17, 18, 19, 21, 22, 23, 25, 26, 27, 31, 32, 35, 36, 39, 41, 45, 46, 47, 52, 54, 56, 57, 60, 62, 63, 64, 65, 66, 67, 68
CoreNet	2000	28	0, 2, 3, 7, 8, 9, 12, 14, 15, 18, 19, 21, 23, 26, 28, 31, 35, 46, 56, 57, 59, 60, 62, 64, 65, 66, 67, 68
CoreNet	2400	17	0, 2, 3, 7, 9, 14, 18, 19, 21, 26, 28, 31, 32, 52, 56, 60, 65

7. Conclusions

In this paper, we propose two heuristic strategies to solve the regenerator placement problem (RPP) in translucent networks under dynamic traffic. The heuristics are based on both traffic forecasts and estimates of blocking probabilities in the network. The proposed heuristics are compared to other regenerator placement algorithms from the literature in two different typologies. We demonstrate the effectiveness of the new proposal under several different situations, such as uniform and non-uniform traffic, fixed and fixed alternative routing algorithms, different values of maximum transmission reach, as well as considering differences between forecast and actual network traffic.

From the obtained results, we conclude that the proposed DLD algorithm finds translucent networks with significant lower BP values than the other investigated algorithms in the paper for the same number of installed regenerators; regardless of the topology, traffic pattern, or intensity considered, transmission reach assumed and whether the fixed of fixed alternate routing is used during the network operation. The superior performance of the DLD algorithm over FLD can be explained by the fact that the DLD uses a more precise node load estimate.

On the other hand, the relative performance of the proposed FLD heuristic and the previously proposed RR, RO, ND, and UN heuristics are very dependent on the considered scenario as the algorithm that achieves the lowest blocking probability values among them is strongly dependent on the topology, maximum transmission reach, number of regenerators deployed, and routing algorithm.

We also believe that the proposed algorithms/methodology can be extended for multi-rate/flex-grid networks by considering, for each rate, it is respective offered load and transmission reach as proposed in Appendix A.

Author Contributions: Conceptualization, A.L.S.d.F. and D.A.R.C.; Investigation, A.L.S.d.F. and D.A.R.C.; Methodology, A.L.S.d.F. and D.A.R.C.; Software, A.L.S.d.F. and D.A.R.C.; Validation, A.L.S.d.F., R.C.A.J. and D.A.R.C.; Writing—original draft, A.L.S.d.F., R.C.A.J. and D.A.R.C.; Writing—review & editing, A.L.S.d.F., R.C.A.J. and D.A.R.C. All authors have read and agreed to the published version of the manuscript.

Funding: This research was partially funded by Government Brazilian researching agencies National Council for Scientific and Technological Development (CNPq) and Fundação de Amparo à Ciência e Tecnologia de Pernambuco (Facepe).

Acknowledgments: The authors acknowledge the National Council for Scientific and Technological Development—CNPq, Facepe, University of Pernambuco, and Federal University of Pernambuco.

Conflicts of Interest: The authors declare no conflict of interest.

Appendix A. Model Extension for Multi-Rate (Elastic) Networks

The formulation is shown in Sections 3 and 4 stands for networks under single bit-rate demands. We can also extend the formulation for a scenario with heterogeneous bit-rate demands. This is the case with elastic optical networks (EON) [11,34,35]. The extension starts by including an extra b sub-index to represent each possible transmission rate in the following symbols: $\omega_{ijb}, \gamma_{ijb}, D_{ijb}, C_{ijb}, K_{ijb}, \Gamma_{nb}$. The set U of allowable B bit rates is $U = \{u_1, u_2, \ldots, u_b, \ldots, u_B\}$. For instance, ω_{ijb} represents the offered load of requests between the nodes i and j that requires a u_b transmission rate. To simplify the notation, we assume that the b-indexed transmission bit rate requires b regenerators to perform its regeneration. This is a valid consideration assuming the virtualized shared pool of elastic regenerators strategy proposed by Jinno et al. [34]. The elastic regenerators consist of an array of fixed-rate spectrum-selective subchannel regenerators (SSRs) that act as a pool of virtualized regeneration resources [34]. Each SSR operates under a fixed transmission bit rate.

Then, we use the theory of erlang loss systems under a multi-slot traffic [31] to evaluate the counterpart of the erlang-b formula under this multi-rate traffic scenario that occurs in EONs. Let us assume a node with s virtualized regenerators. A state $e = (I_1, I_2, \ldots, I_B)$ represents the number of active u_b-rate connections using SSRs in such a node, where I_b is the number of active u_b-rate connections, each of which, as explained before, using b regeneration resources. All possible states when s SSRs are installed in a node form the set $\mathcal{E}_s$, which includes any occurrence of values for $(I_1, I_2, \ldots, I_B)$ so that $\sum_{b=1}^{B} b \cdot I_b \leq s$. Let $\mathcal{E}_s(b)$ be the set of states that cannot admit a u_b-rate demand, i.e., $\mathcal{E}_s(b) = \left\{\forall e | \sum_{x=1}^{B} x \cdot I_x > s - b\right\}$, as the summation accounts for the number of SSRs in use in the analyzed node and $s - b$ accounts for the maximum number of SSRs in use in the node that still admits a regeneration of a u_b-rate demand in such node. In such scenario, we can replace (3) and (4) by

$$\Gamma_{nb} = \sum_{i=1}^{N} \sum_{j=1}^{N} E_{ijb}(n), \tag{A1}$$

in which E_{ijb} is a place holder for either $G_{ijb}(n) = \gamma_{ijb}$ if $\exists\, m \,/\, n \in c_{ijb}^{(m)}$ and 0 otherwise or for F_{ijb}:

$$F_{ijb}(n) = \begin{cases} \frac{\gamma_{ijb}}{M'_{ij}} \sum_{m=1}^{M'_{ij}} P_{ijm}(n), & \text{if } n \in \pi_{ij} \\ 0, & \text{otherwise,} \end{cases} \tag{A2}$$

with $P_{ijm}(n) = 1$ if $n \in d_{ijb}^{(m)}$ and 0 otherwise.

In a node n with s SSRs, the probability of a state $e \in \mathcal{E}_s$ is given by [31]

$$p_n(e,s) = Q_s \prod_{b=1}^{B} \frac{\Gamma_{nb}^{I_b}}{I_b!}, \tag{A3}$$

in which Γ_{nb} is the total offered load estimate from u_b-rate demands on node n and Q_s is the normalization coefficient given by [31]

$$Q_s = \frac{1}{\sum\limits_{e \in \mathcal{E}_s} p_n(e,s)}. \tag{A4}$$

Then, the blocking probability of u_b-rate demands in node n may be evaluated by

$$BP_n(s,b) = \sum_{e \in \mathcal{E}_s(b)} p_n(e,s), \tag{A5}$$

and (6) can be replaced by

$$BP_N(n,s) = \frac{\sum_{i=1}^{N} \sum_{j=1}^{N} \sum_{b=1}^{B} PB_n(s,b) \cdot E_{ijb}(n)}{\sum_{i=1}^{N} \sum_{j=1}^{N} \sum_{b=1}^{B} \omega_{ijb}}. \tag{A6}$$

For a node n with a total of $r(n)$ installed regenerators, it is possible to replace (8) in a multi-rate scenario using

$$\Delta BP_N(n,r(n)) = BP_N(n,r(n)) - BP_N(n,r(n)+1). \tag{A7}$$

References

1. Yang, X.; Ramamurthy, B. Sparse Regeneration in Translucent Wavelength-Routed Optical Networks: Architecture, Network Design and Wavelength Routing. *Photonic Netw. Commun.* **2005**, *10*, 39–53. [CrossRef]
2. Pedro, J. Predeployment of regenerators for fast service provisioning in DWDM transport networks [Invited]. *J. Opt. Commun. Netw.* **2015**, *7*, A190–A199. [CrossRef]
3. Nath, I.; Chatterjee, M.; Bhattacharya, U. A survey on regenerator Placement Problem in translucent optical network. In Proceedings of the 2014 International Conference on Circuits, Systems, Communication and Information Technology Applications (CSCITA), Mumbai, India, 4–5 April 2014; pp. 408–413. [CrossRef]
4. Chaves, D.A.R.; Carvalho, R.V.B.; Pereira, H.A.; Bastos-Filho, C.J.A.; Martins-Filho, J.F. Novel strategies for sparse regenerator placement in translucent optical networks. *Photonic Netw. Commun.* **2012**, *24*, 237–251. [CrossRef]
5. da Silva, E.; Almeida, R.C., Jr.; Pereira, H.; Chaves, D. Assessment of novel regenerator assignment strategies in dynamic translucent elastic optical networks. *Photonic Netw. Commun.* **2020**, *39*, 54–69. [CrossRef]
6. Nath, I.; Bhattacharya, U.; Chatterjee, M. An Efficient Heuristic to Minimize Number of Regenerations in Translucent Optical Network under Dynamic Scenario. In Proceedings of the 20th International Conference on Distributed Computing and Networking, Bangalore, India, 4–7 January 2019; pp. 451–454. [CrossRef]
7. Chaves, D.A.R.; da Silva, E.F.; Bastos-Filho, C.J.A.; Pereira, H.A.; Almeida, R.C. Heuristic algorithms for regenerator assignment in dynamic translucent elastic optical networks. In Proceedings of the 2015 17th International Conference on Transparent Optical Networks (ICTON), Budapest, Hungary, 5–9 July 2015; pp. 1–4. [CrossRef]
8. De Groote, M.; Manousakis, K.; Kokkinos, P.; Colle, D.; Pickavet, M.; Christodoulopoulos, K.; Varvarigos, E.; Demeester, P. Cost comparison of different translucent optical network architectures. In Proceedings of the 2010 9th Conference of Telecommunication, Media and Internet, Ghent, Belgium, 7–9 June 2010; pp. 1–8. [CrossRef]
9. Rahman, Q.; Bandyopadhyay, S.; Aneja, Y. Optimal regenerator placement in translucent optical networks. *Opt. Switch. Netw.* **2015**, *15*, 134–147. [CrossRef]
10. Leitner, M.; Ljubić, I.; Riedler, M.; Ruthmair, M. Exact Approaches for Network Design Problems with Relays. *INFORMS J. Comput.* **2019**, *31*, 171–192. [CrossRef]

11. Cavalcante, M.A.; Pereira, H.A.; Chaves, D.A.R.; Almeida, R.C. Evolutionary Multiobjective Strategy for Regenerator Placement in Elastic Optical Networks. *IEEE Trans. Commun.* **2018**, *66*, 3583–3596. [CrossRef]
12. Aibin, M.; Walkowiak, K. Regenerator placement algorithms for cloud-ready Elastic Optical Networks. In Proceedings of the 2015 17th International Conference on Transparent Optical Networks (ICTON), Budapest, Hungary, 5–9 July 2015; pp. 1–4. [CrossRef]
13. Chaves, D.A.R.; Ayres, C.F.C.L.C.; Carvalho, R.V.B.; Pereira, H.A.; Bastos-Filho, C.J.A.; Martins-Filho, J.F. Multiobjective sparse regeneration placement algorithm in optical networks considering network performance and CAPEX. In Proceedings of the 2010 12th International Conference on Transparent Optical Networks, Munich, Germany, 27 June–1 July 2010; pp. 1–4. [CrossRef]
14. Rizzelli, G.; Musumeci, F.; Tornatore, M.; Maier, G.; Pattavina, A. Wavelength-aware translucent network design. In Proceedings of the 2011 Optical Fiber Communication Conference and Exposition and the National Fiber Optic Engineers Conference, Los Angeles, CA, USA, 6–10 March 2011; pp. 1–3.
15. Agarwal, M.; Chatterjee, M.; Chandran, S. Offline routing and wavelength assignment for identification of regeneration sites in translucent WDM optical networks. In Proceedings of the 2017 International Conference on Advances in Computing, Communications and Informatics (ICACCI), Udupi, India, 13–16 September 2017; pp. 623–628. [CrossRef]
16. Li, X.; Aneja, Y. Regenerator location problem: Polyhedral study and effective branch-and-cut algorithms. *Eur. J. Oper. Res.* **2017**, *257*, 25–40. [CrossRef]
17. Mertzios, G.; Shalom, M.; Wong, P.; Zaks, S. Online Regenerator Placement. *Theory Comput. Syst.* **2017**, *61*, 739–754. [CrossRef]
18. Yan, L.; Xu, Y.; Brandt-Pearce, M.; Dharmaweera, N.; Agrell, E. Robust regenerator allocation in nonlinear flexible-grid optical networks with time-varying data rates. *IEEE/OSA J. Opt. Commun. Netw.* **2018**, *10*, 823–831. [CrossRef]
19. Li, X.; Aneja, Y. A new branch-and-cut approach for the generalized regenerator location problem. *Ann. Oper. Res.* **2020**, *295*, 229–255. [CrossRef]
20. Li, X.; Aneja, Y. A Branch-And-Benders-Cut approach for the fault tolerant regenerator location problem. *Comput. Oper. Res.* **2020**, *115*, 104847. [CrossRef]
21. Zhao, J.; Subramaniam, S.; Brandt-Pearce, M. Efficient and accurate analytical performance models for translucent optical networks. *J. Opt. Commun. Netw.* **2014**, *6*, 1128–1142. [CrossRef]
22. Chaves, D.A.R.; Cavalcante, M.A.; Pereira, H.A.; Almeida, R.C. A case study of regenerator placement and regenerator assignment in dynamic translucent elastic optical networks. In Proceedings of the 2016 18th International Conference on Transparent Optical Networks (ICTON), Trento, Italy, 10–14 July 2016; pp. 1–4. [CrossRef]
23. Walkowiak, K.; Klinkowski, M. Energy Efficiency of Dynamic Routing in Elastic Optical Networks with Back-to-Back Regeneration. In Proceedings of the 2018 International Conference on Computing, Networking and Communications (ICNC), Maui, HI, USA, 5–8 March 2018; pp. 399–403. [CrossRef]
24. Yang, H.; Zhao, X.; Yao, Q.; Yu, A.; Zhang, J.; Ji, Y. Accurate Fault Location using Deep Neural Evolution Network in Cloud Data Center Interconnection. *IEEE Trans. Cloud Comput.* **2020**. [CrossRef]
25. Leland, W.E.; Taqqu, M.S.; Willinger, W.; Wilson, D.V. On the self-similar nature of Ethernet traffic (extended version). *IEEE/ACM Trans. Netw.* **1994**, *2*, 1–15. [CrossRef]
26. Karagiannis, T.; Molle, M.; Faloutsos, M.; Broido, A. A nonstationary Poisson view of Internet traffic. In Proceedings of the IEEE INFOCOM 2004, Hong Kong, China, 7–11 March 2004; Volume 3, pp. 1558–1569. [CrossRef]
27. Terdik, G.; Gyires, T. Does the Internet Still Demonstrate Fractal Nature? In Proceedings of the 2009 Eighth International Conference on Networks, Guadeloupe, France, 1–6 March 2009; pp. 30–34. [CrossRef]
28. Terdik, G.; Gyires, T. LEvy Flights and Fractal Modeling of Internet Traffic. *IEEE/ACM Trans. Netw.* **2009**, *17*, 120–129. [CrossRef]
29. Walkowiak, K.; Klinkowski, M.; Włodarczyk, A.; Kasprzak, A. Predeployment of transponders for dynamic lightpath provisioning in translucent spectrally-spatially flexible optical networks. *Appl. Sci.* **2020**, *10*, 2802. [CrossRef]
30. Cheng, S.; Xiao, D.; Huang, A.; Aibin, M. Machine learning for regenerator placement based on the features of the optical network. In Proceedings of the 2019 21st International Conference on Transparent Optical Networks (ICTON), Angers, France, 9–13 July 2019. [CrossRef]
31. Anurag Kumar, D.M.; Kuri, J. *Communication Networking: An Analytical Approach*, 1st ed.; Morgan Kaufmann: Burlington, MA, USA, 2004.
32. DARPA. Core Optical Networks (CORONET) Continental United States (CONUS) Topology. Available online: http://monarchna.com/CORONET_CONUS_Topology.xls (accessed on 1 January 2018).
33. Cavalcante, M.; Pereira, H.; Chaves, D.; Almeida, R. Optimizing the cost function of power series routing algorithm for transparent elastic optical networks. *Opt. Switch. Netw.* **2018**, *29*, 57–64. [CrossRef]
34. Jinno, M.; Takara, H.; Yonenaga, K.; Hirano, A. Virtualization in optical networks from network level to hardware level. *J. Opt. Commun. Netw.* **2013**, *5*, A46–A56. [CrossRef]
35. Yang, H.; Yao, Q.; Bao, B.; Yu, A.; Zhang, J.; Vasilakos, A.V. Multi-Associated Parameters Aggregation-Based Routing and Resources Allocation in Multi-Core Elastic Optical Networks. *IEEE/ACM Trans. Netw.* **2022**, 1–13. [CrossRef]

applied sciences

MDPI

Communication

200 Gbps/λ PON Downstream C-Band Direct-Detection Links with ≥29 dB Power Budget

Haoyi Wang [1], Pablo Torres-Ferrera [1], Giuseppe Rizzelli [1,2,*], Roberto Mercinelli [3], Valter Ferrero [1] and Roberto Gaudino [1]

1 Department of Electronics and Telecommunications, Politecnico di Torino, Corsso Duca degli Abruzzi 24, 10129 Torino, Italy; haoyi.wang@polito.it (H.W.); pablo.torres@polito.it (P.T.-F.); valter.ferrero@polito.it (V.F.); roberto.gaudino@polito.it (R.G.)
2 LINKS Foundation, Via Pier Cario Boggio 61, 10138 Torino, Italy
3 Technology Innovation Cloud Access Solutions, Telecom Italia (TIM), Via Reiss Romoli 274, 10148 Torino, Italy; roberto.mercinelli@telecomitalia.it
* Correspondence: giuseppe.rizzelli@linksfoundation.com

Abstract: In this paper we present the simulative analysis of a 200 Gbps per wavelength (λ) 8-level pulse amplitude modulation (PAM-8) downstream communication over up to 20 km single mode fiber (SMF) in C-band based on direct detection (DD) achieving at least a 29 dB link power budget in a PON environment. We use chromatic dispersion digital pre-compensation (CD-DPC) and a dual-arm in-phase and quadrature Mach–Zehnder modulator (IQ-MZM) at the optical line termination (OLT) side, while preserving DD in the optical network unit (ONU). Three receiver digital-signal-processing (DSP) options are analyzed and compared: square-root-like technique (SQRT) in combination with a feed forward equalizer (FFE) and a decision feedback equalizer (DFE), the Volterra nonlinear equalizer (VNLE), and the SQRT in combination with the VNLE. The SQRT can be applied in combination with the VNLE to decrease the receiver DSP complexity while maintaining the required system performance. We show that PAM-8 with CD-DPC and the SQRT in combination with the VNLE is a feasible solution for 200 Gbps per λ downstream C-band transmission for PON.

Keywords: PON; C-band; chromatic dispersion compensation; direct detection; 200 Gbps per wavelength

Citation: Wang, H.; Torres-Ferrera, P.; Rizzelli, G.; Mercinelli, R.; Ferrero, V.; Gaudino, R. 200 Gbps/λ PON Downstream C-Band Direct-Detection Links with ≥29 dB Power Budget. *Appl. Sci.* **2022**, *12*, 3538. https://doi.org/10.3390/app12073538

Academic Editor: John Xiupu Zhang

Received: 25 February 2022
Accepted: 29 March 2022
Published: 30 March 2022

1. Introduction

Next-generation passive optical networks (PON) have been under rapid development to fulfill the ever-increasing bandwidth and traffic requirements in the access and short-reach communications segment. The standardization of higher speed PON targeting 50 Gbps per wavelength (λ) has been recently performed by ITU-T [1–4]. However, next-generation 5G and 6G systems will require even higher data rates, and PONs represent a suitable practical technology to support future optical access solutions [5–8]. Currently, research efforts are targeting data rates of 100 Gbps per λ and beyond [9–17] for the next step of PON evolution.

The latest PON standards (i.e., 25G-PON and 50G-PON) have chosen the direct detection (DD) scheme to meet cost requirements, and O-band operations for both upstream (US) and downstream (DS) directions to cope with chromatic dispersion [1–4]. However, the penalty due to chromatic dispersion is not negligible at high data rates, such as 200 Gbps, even in the O-band. Compared to the O-band, the C-band can provide a lower optical loss and optical nonlinearity. For higher speed PON upgrades, backward compatibility with legacy Gigabit PON (G-PON) (currently, one of the most commonly deployed PON standards) can be achieved in the C-band by reusing the 1480–1550 nm wavelength range of the downstream. Moreover, the O-band window might become filled, opening to the possibility of considering the C-band again in future PON standardization efforts. To achieve this, dispersion must be compensated. Coherent detection is well-known to allow compensation of CD, power-fading effect and other impairments through digital signal processing, thus enabling C-band operation. In [18–20], the possibility of using coherent detection has been studied. However, due to the high cost and complexity, the use of coherent detection in the cost-effective PON environment is still under discussion [9,18]. In a recent work [15], we proposed chromatic dispersion digital pre-compensation (CD-DPC) for 100 Gbps links using a dual-arm in-phase and quadrature Mach–Zehnder modulator (IQ-MZM) at the transmitter (TX) side while keeping DD at the receiver (RX). This approach only increases complexity at the optical line termination (OLT) side, which in the PON downstream scenario is shared among the end users.

Increasing the target data rate requires pushing the system into nonlinear operation [16,17,21] to be able to increase the transmitter launch power and the optical modulation index to achieve the demanding power–budget requirements. In [16], we analyzed simple nonlinear compensation techniques, i.e., square-rootlike (SQRT) and polynomial (POLY) techniques. We simulatively and experimentally confirmed about 2 dB gain in terms of maximum optical distribution network (ODN) loss for a 100 Gbps downstream transmission, thanks to SQRT and POLY.

Next generation PONs must be backward compatible with legacy PONs, and the ODN must remain the same. Typically, at least a 29 dB link power budget (budget class N1) and a 0 to (at least) 20 km fiber reach must be guaranteed [1–4,22]. Regarding the trend of ITU-T PON standardization evolution, at least a 4-fold increment in DS capacity can be observed between 2 consecutive PON system generations [2,4]. For example, there is a 4-fold increment in DS capacity from a GPON (with DS 2.5 Gbps per λ) to XG(S)-PON (with DS 10 Gbps per λ), and a 5-fold increment from a XG(S)-PON to a higher speed PON (with DS 50 Gbps per λ). Therefore, it is expected that the next generation beyond 50G-PON will target (at least) 200 Gbps per λ in DS. To meet the 29 dB ODN loss and 20 km fiber reach at 100 Gbps is very challenging, as we showed in previous works [14–17]. For a 200 Gbps per λ transmission, the situation becomes extremely critical. In this paper, we focus on 200 Gbps per λ transmission in a C-band to guarantee 29 dB ODN loss over a 20 km fiber link, by combining the CD-DPC algorithm with two nonlinear distortion compensation techniques: the SQRT [16] and the Volterra nonlinear equalizer (VNLE) [17]. In [17], for 100 Gbps per λ transmission in a C-band, we experimentally showed a 1–2 dB gain in term of ODN loss by using VNLE when compared to the SQRT.

Until now, non-return-to-zero on-off-keying (NRZ-OOK) modulation format has been defined for all standardized PONs. It is expected that 50G class optoelectronics (O/E) will be commercially available by the time 200 Gbps products are developed. However, devices to support OOK and 4-level pulse amplitude modulation (PAM-4) operations at 100 Gbaud or beyond seems difficult to achieve in the midterm [3,23]. Thus, we consider 8-level PAM (PAM-8) as a candidate modulation format for 200 Gbps transmission to have a reduced baud rate (i.e., around 66.67 Gbaud for 200 Gbps bit rate). In [13], a 200 Gbps PAM-4 communication with a 29 dB ODN loss was experimentally demonstrated by using directly modulated laser (DML), direct detection and Raman amplification over a 21 km fiber. However, very high transmitter and receiver bandwidth devices, i.e., a 65 GHz DML and 70 GHz P-type-intrinsic-N-type diodes (PIN), were used in the experiments. Moreover, the strong Raman pumping used in the experiments is not suited for a cost-effective short-reach PON. In this paper, we demonstrate over 29 dB of ODN loss using 50G O/E devices without Raman amplification. In a 50G PON, the maximum launched power into ODN is +7, +9, +11 and +11 dBm for budget class N1, N2, E1 and E2, respectively [24]. For 50G PON with a higher budget class, for example Class C+ (32 dB) and beyond, a minimum launched power of +10 dBm might be required. In PONs, a launch power a few dB higher than the minimum one can be tolerated [25]. Transmitted optical power (TOP) could be increased to achieve the required link power budget. An electro absorption modulated laser (EML) boosted by an external semiconductor optical amplifier (SOA) can be placed at the TX side to increase launch power in PONs for a wide wavelength range of 1200–1650 nm [26]. For a 50G PON, EML + SOA was exploited DS to support high budget classes. In [25], a 14 dBm launch power was emulated by using EML + SOA. In [27], an integrated EML + SOA with a 13 dBm output power was experimentally demonstrated to achieve a 35 dB link budget. In our previous works for 100 Gbps transmission, we used a TOP of 11 dBm. In this paper, we increase it to 15 dBm (when the VNLE is applied) to achieve the 29 dB power–budget requirements for 200 Gbps transmission.

The paper is organized as follows: the details of the simulation setup and simulated parameters are described in Section 2. In Section 3, we analyze the proposed system performance through simulations with variable transceiver bandwidth and for different RX DSP options. The complexity of different RX DSP options is compared in terms of multiplication per sample (MPS). Lastly, we discuss the results and draw some conclusions in Section 4.

2. Simulation Setup

The simulation setup is schematically illustrated in Figure 1. Single channel transmission is considered. At the TX side, a PRBS-15 pseudorandom binary sequence is generated at bit rate R_b = 200 Gbps, and then coded to generate a PAM-8 sequence. As shown in Figure 1, the CD-DPC algorithm is implemented based on a finite impulse response (FIR) complex filter, which is explained in detail in [15]. The key idea behind this algorithm is the following: the fiber length L is known in advance and the accumulated CD is $D{\cdot}L$ ps/nm (where D is the dispersion coefficient). An electrical transmitted signal is sent to the FIR filter-based CD-DPC block, which works at m samples per symbol (SpS) with N_t taps. The CD-DPC block is placed at the TX side digital signal processing (DSP) and emulates the inverse of the CD accumulated along the link. A proper value of pre-compensated length L_C must be set in the CD-DPC block to design the taps of the FIR filter. As a result, an accumulated CD of $D{\cdot}L_C$ ps/nm is assumed to be pre-compensated. There could be a mismatch ΔL between L and L_C, ($\Delta L = L_C - L$), which the system should tolerate. In [14–17], we demonstrated a tolerance on ΔL of about $\pm$2 km in a C-band for a 100 Gbps transmission through simulations and experiments. In this paper, the CD-DPC DD works at m = 2 SpS and N_t = 80. The I and Q arm of the dual-arm IQ-MZM are driven by two real-valued in-phase $I(n)$ and quadrature $Q(n)$ signals biased at quadrature and null point, respectively. The peak-to-peak amplitude V_{PP} of signal $I(n)$ and $Q(n)$, i.e., $V_{PP,I}$ and $V_{PP,Q}$ respectively, are measured at the input of the IQ-MZM. $V_{PP,I}$ and $V_{PP,Q}$ must be optimized to obtain the best system performance by optimizing a scaling factor (the parameter k in Figure 1, which is the same for both $I(n)$ and $Q(n)$ signal). The simulator is explained in detail in [15]. The bias voltage of the IQ-MZM V_π is 5 V.

The modulated optical signal in C-band at wavelength λ = 1550 nm is propagated over a conventional ITU-T G.652 single mode fiber (SMF) with length L = 20 km (typical for PON). The fiber attenuation, chromatic dispersion D and non- linear coefficient n_2 are set to 0.22 dB/km, 17 ps/(nm·km) and $26{\times}10^{-21}$ m^2/W, respectively. The effective area is 80 µm^2 [28]. Kerr nonlinearities are introduced using the conventional nonlinear Schrödinger equation (NLSE) solved numerically by the split-step Fourier method. The received optical power (ROP) is measured after a variable optical attenuator (VOA) is used to emulate the ODN loss. Second order low-pass super Gaussian filters (SGF) are used to emulate the O/E bandwidth limitations with the same −3 dB bandwidth f_{3dB} both at TX and RX side. We analyze the impact of transceiver bandwidth limitations varying f_{3dB} from 30% to more than 100% of the baud rate, i.e., from 20 GHz to 70 GHz. At the RX side, an optical filter with a 75 GHz bandwidth is placed between the SOA and PIN. The receiver simulated parameters are SOA gain G = 11 dB, SOA noise figure NF = 7 dB, photodiode responsivity R = 0.7 A/W, and transimpedance amplifier (TIA) input referred noise density $IRND$ = 12 pA/sqrt(Hz). A digital-to-analog converter (DAC) and an analog-to-digital converter (ADC) with a 6 bits resolution for quantization are used to emulate the arbitrary waveform generator (AWG) and real-time oscilloscope (RTO), respectively.

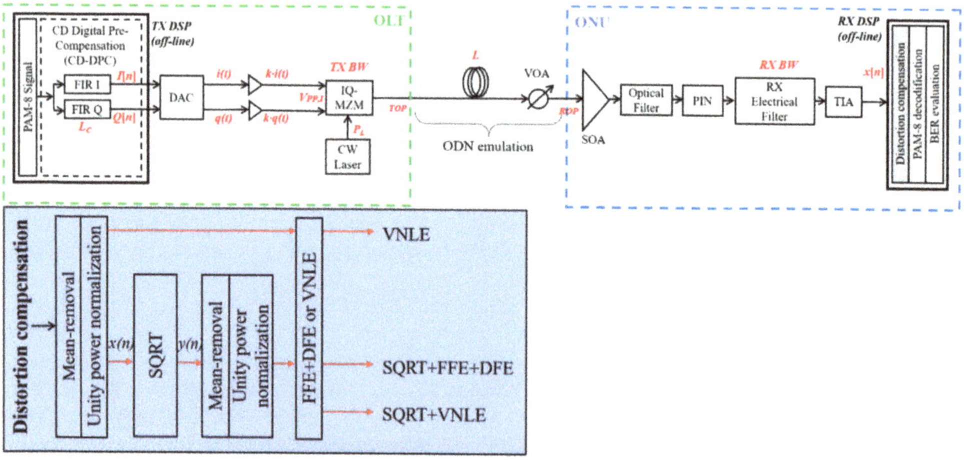

Figure 1. Simulation setup of the proposed CD-DPC DD link, emulating a PON downstream transmission. Dispersion compensation blocks with different DSP options are indicated.

In the off-line DSP at the receiver, the digitized sequence is down sampled and off-line processed at 2 SpS, using one of the three DSP nonlinear distortion compensation methods shown at the bottom of Figure 1. As detailed in [16], the SQRT exploits a square-rootlike function, i.e., $y(n) = \sqrt{x(n) + |min(x(n))|}$, and POLY consists of a polynomial function, i.e., $y(n) = x(n) + \alpha x^2(n)$, where $x(n)$ and $y(n)$ are the input and output signal of the SQRT (or POLY) block as shown in Figure 1, and α is a free parameter to be optimized. Since we also showed that a SQRT can provide a similar gain as POLY, but with lower complexity, in this paper we will consider the SQRT. We also analyze a technique based on the VNLE. To keep the complexity low, the VNLE is limited to third order, i.e., linear, quadratic, and cubic terms are considered [29]. In our simulations, we set quadratic memory and cubic memory to be the same, and term them as nonlinear memory (NL memory). For example, *NL memory* = 5 samples means *quadratic memory* = *cubic memory* = 5 samples. The three DSP nonlinear distortion compensation methods are: (i) the SQRT [16,30] in combination with a 120-taps feed forward equalizer (FFE) and a 5 taps decision feedback equalizer (DFE). We indicate it as "SQRT + FFE + DFE". (ii) VNLE with 121 linear memory and a variable number of NL memory, ranging from 5 samples to 20 samples. We indicate it as "VNLE". (iii) The SQRT in combination with the same VNLE as in option (ii). We indicate it as "SQRT + VNLE". The SQRT + FFE + DFE method is equivalent to the VNLE method with *NL memory* = 0. The transmitted optical power *TOP* is set to 15 dBm for options with the VNLE (i.e., VNLE and SQRT + VNLE) and 11 dBm for the option without the VNLE (i.e., SQRT + FFE + DFE).

The typical metric used in a PON performance evaluation is the achievable ODN loss, calculated as the ratio between the TOP and the ROP. In this manuscript, we consider the required ROP (RROP) to obtain a BER target $BER_T = 10^{-2}$ (low-density parity check (LDPC) code forward error correction (FEC) scheme), and we will show system performance in terms of the maximum ODN loss that that guarantees operation equal to or below the BER_T (*Maximum ODN loss [dB] = TOP [dBm] − RROP [dBm]*).

3. Simulation Results

In our simulation setup, shown in Figure 1, we varied the transceiver bandwidth up to 70 GHz and the VNLE NL memory from 0 to 20 samples. We started by showing the impact of the normalized peak-to-peak amplitude of the in-phase signal $V_{PP,I}/V_\pi$ by optimizing $V_{PP,I}$ for options VNLE and SQRT + VNLE. Figures 2–5 shows the system performance when L = 20 km and *TOP* = 15 dBm, in terms of the maximum ODN loss as a function of TX and RX bandwidth and of the normalized amplitude $V_{PP,I}/V_{\pi}$, for *NL memory* = 5 samples, 10 samples, 15 samples and 20 samples, respectively. Note that the pre-compensated length L_C is optimized for each pair of the two parameters. The optimum L_C is slightly smaller than the fiber length L, and the difference ΔL between L and L_C increases with TOP. For instance, L_C is around 18 km when L = 20 km and *TOP* = 15 dBm. This phenomenon is due to the interaction between the fiber Kerr effect and the chromatic dispersion, as explained in [15,31].

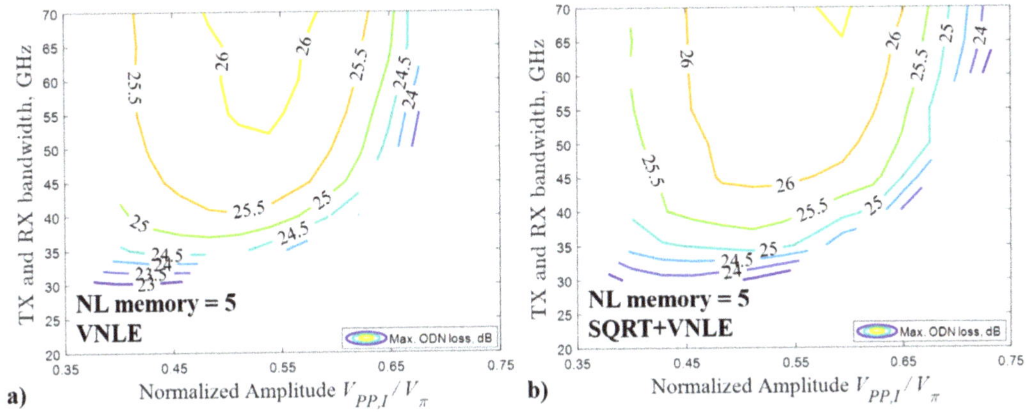

Figure 2. Maximum ODN loss at $BER_T = 10^{-2}$ as a function of TX and RX bandwidth and of the normalized peak-to-peak amplitude of the in-phase signal $V_{PP,I}/V_{\pi}$ with *NL memory* = 5 samples. The fiber length is L = 20 km. The transmitted optical power is *TOP* = 15 dBm: (**a**) VNLE; and (**b**) SQRT + VNLE.

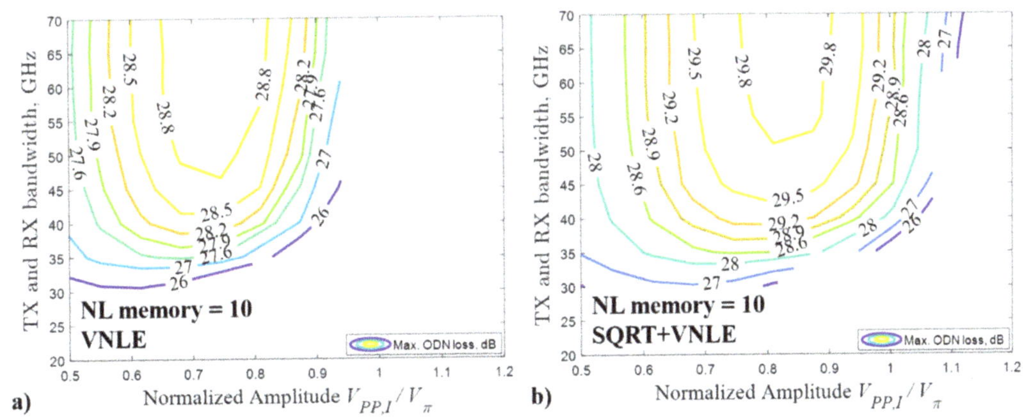

Figure 3. Maximum ODN loss at $BER_T = 10^{-2}$ as a function of TX and RX bandwidth and of the normalized peak-to-peak amplitude of the in-phase signal $V_{PP,I}/V_{\pi}$ with *NL memory* = 10 samples. The fiber length is L = 20 km. The transmitted optical power is *TOP* = 15 dBm: (**a**) VNLE; and (**b**) SQRT + VNLE.

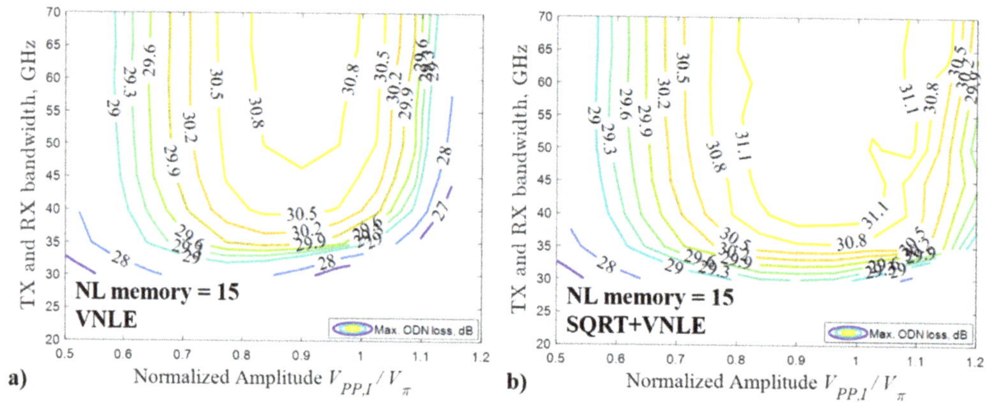

Figure 4. Maximum ODN loss at $BER_T = 10^{-2}$ as a function of TX and RX bandwidth and of the normalized peak-to-peak amplitude of the in-phase signal $V_{PP,I}/V_\pi$ with *NL memory* = 15 samples. The fiber length is L = 20 km. The transmitted optical power is *TOP* = 15 dBm: (**a**) VNLE; and (**b**) SQRT + VNLE.

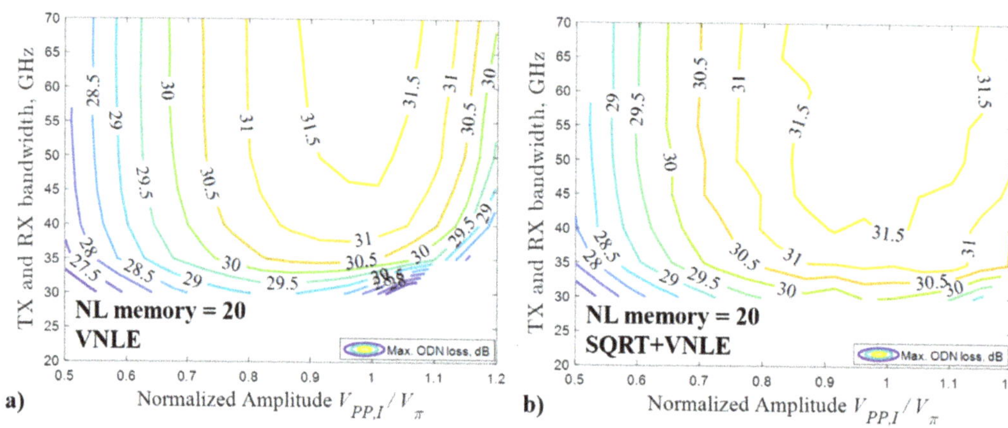

Figure 5. Maximum ODN loss at $BER_T = 10^{-2}$ as a function of TX and RX bandwidth and of the normalized peak-to-peak amplitude of the in-phase signal $V_{PP,I}/V_\pi$ with *NL memory* = 20 samples. The fiber length is L = 20 km. The transmitted optical power is *TOP* = 15 dBm: (**a**) VNLE; and (**b**) SQRT + VNLE.

From Figures 2–5, we can observe that the optimum normalized amplitude $V_{PP,I}/V_\pi$ is in the range 0.55–0.60, 0.70–0.90, 0.90–1.00 and 1.00–1.10 for the VNLE NL memory equal to 5 samples, 10 samples, 15 samples and 20 samples, respectively. The optimum normalized amplitude $V_{PP,I}/V_\pi$ increases as the NL memory increases. This can be explained considering the effect of the optical modulation index (OMI) on the nonlinearities generated in the modulation process. The OMI can be used to find the optimum operation point (at which the highest modulation levels can be obtained without introducing additional distortions) [32]. Smaller $V_{PP,I}/V_\pi$ results in lower nonlinearity of the modulation, but also into a lower modulation level. In contrast, larger $V_{PP,I}/V_\pi$ can result in a higher modulation level but also in stronger nonlinear effects of the modulation. Therefore, it is very important to set the optimum driving signal amplitude by optimizing the normalized amplitude $V_{PP,I}/V_\pi$. Without TX and RX bandwidth limitations and at the optimum $V_{PP,I}/V_\pi$, when the VNLE is applied alone, the maximum ODN loss of about 26.1 dBm,

29.1 dBm, 31.0 dBm, and 31.7 dBm can be reached for *NL memory* = 5 samples, 10 samples, 15 samples and 20 samples, respectively; when SQRT + VNLE is applied, the maximum ODN loss of about 26.6 dBm, 30.0 dBm, 31.7 dBm, and 32.0 dBm can be reached for *NL memory* = 5 samples, 10 samples, 15 samples and 20 samples, respectively. The VNLE can provide up to 3 dB improvement for every five samples additional NL memory, while the SQRT can provide about 0.5–1.0 dB improvement compared to the VNLE only option. However, the complexity of the VNLE is much higher than that of the SQRT (the complexity will be discussed and analyzed in detail in the following section). It is a trade-off between the system performance and the complexity of RX DSP options.

In addition to the modulation nonlinear effect and the fiber Kerr nonlinear effect, the nonlinear effect in the electrical domain caused by chromatic dispersion after the modulus square operation of DD (DD is inherently nonlinear due to the square-law detection) can be mitigated by using the SQRT as shown in [30]. By comparing Figures 2–5, we can also notice that the optimum normalized amplitude $V_{PP,I}/V_{\pi}$ of the option SQRT + VNLE is slightly larger than that of the VNLE method (with the same NL memory). Therefore, a higher OMI can be set using VNLE with high NL memory combined with the SQRT, which allows to compensate for stronger nonlinear distortion.

Figure 6 shows the maximum ODN loss as a function of TX and RX bandwidth and of the VNLE NL memory at the optimum normalized amplitude $V_{PP,I}/V_{\pi}$, for 20 km SMF. On the x-axis, the VNLE *NL memory* equal to 0 corresponds to the SQRT + FFE + DFE technique used in the receiver DSP. This option, which does not include the VNLE, is used as the lowest complexity reference case. Note that, in all cases, the transmitted optical power, the normalized amplitude $V_{PP,I}/V_{\pi}$ and the pre-compensated length L_C are optimized. The optimum *TOP* is 11 dBm and 15 dBm for the SQRT + FFE + DFE and VNLE techniques, respectively. We limit our analysis to TOP = 15 dBm as higher transmitted power levels are unrealistic in PON applications. The optimum L_C is 19 km for SQRT + FFE + DFE (with 11 dBm TOP) and 18 km for the VNLE (with 15 dBm TOP). The optimum L_C decreases as TOP increases. This is related to the fact that a higher transmitted optical power results in a higher fiber nonlinear interference [33], which partially counteracts CD due to interaction between the CD and the Kerr effect [15,31]. At the very demanding 200 Gbps rate, the system performance is very poor (maximum ODN loss <23 dB) when strong TX and RX bandwidth limitations are present (i.e., $\leq$25 GHz) even when using the most complex DSP approach (VNLE + SQRT with *NL memory* = 20 samples). For SQRT + FFE + DFE (i.e., *NL memory* = 0), the largest maximum ODN loss is only about 25 dB even if using very high 70 GHz TX and RX bandwidth. Therefore, this simple DSP solution is not enough to meet the ODN loss requirement with PAM-8 200 Gbps transmission. By using the VNLE, both alone and in combination with the SQRT, a considerable gain can be obtained when we increase the NL memory. As shown in Figures 2–6, about 3 dB, 5 dB and 6 dB maximum ODN loss increase can be obtained when the NL memory is increased from 5 samples to 10 samples, 15 samples and 20 samples, respectively. The trend shows that the improvement would be very limited by further increasing the NL memory above 20 samples. This means that almost all the fiber nonlinear interference is compensated for by the VNLE. About a 3 dB gain in terms of maximum ODN loss can be obtained by relaxing the TX and RX bandwidth limitation from 30 GHz to 50 GHz, with the same NL memory. As shown in Figure 6, for a fixed NL memory value, the improvement is more evident when increasing the TX and RX bandwidth from 30 to 40 GHz. When the TX and RX bandwidth are above 50 GHz, only very limited improvement can be obtained when TX and RX bandwidth are increased.

System performance cannot be usually enhanced by relaxing the bandwidth limitations. More powerful RX DSP options are needed. Comparing the VNLE and the SQRT + VNLE, we observe a non-negligible ODN loss gain of about 1 dB, at the cost of slightly increased complexity due to the SQRT. Moreover, in some cases the overall complexity can be reduced by introducing the SQRT. As shown in Figure 6a), in the 35 GHz to 40 GHz TX and RX bandwidth range representing commercially available 50G-O/E [23], the VNLE approach requires NL memory from 12 samples to 13 samples to achieve 29 dB maximum ODN loss, whereas a lower 9 samples to 11 samples NL memory is needed when combining the VNLE and the SQRT (see Figure 6b)). This reduction in NL memory can provide a significant improvement in RX DSP complexity in terms of MPS, which will be discussed later. When *NL memory* = 20 samples is applied, the 29 dB maximum ODN loss can be reached even with 30 GHz TX and RX bandwidth, at the cost of a very high DSP complexity (which will be discussed later).

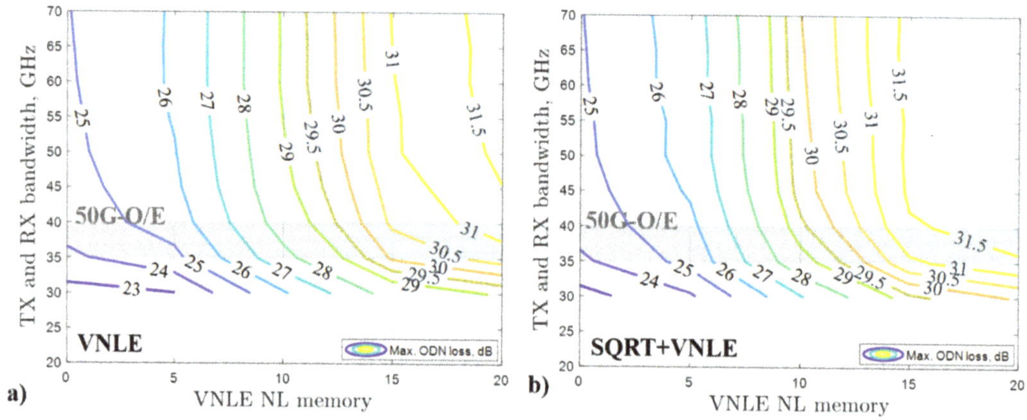

Figure 6. Maximum ODN loss at $BER_T = 10^{-2}$ as a function of TX/RX bandwidth and nonlinear memory of VLNE. The fiber length is L = 20 km: (**a**) VNLE; and (**b**) SQRT + VNLE. The light blue area is where ODN loss is above 29 dB, and the light grey area is where commercial 50G-O/E are considered. The VNLE NL memory = 0 is equivalent to RX DSP option SQRT + FFE + DFE. TOP is 11 dBm for SQRT + FFE + DFE and 15 dBm for all other RX DSP options.

It is typical in PON architectures that each served ONU is located at a different distance from the OLT. In Figure 7, we present the maximum ODN loss as a function of fiber link lengths L ranging from 5 km to 20 km, which is the typical reach range in PONs. TX and RX bandwidth is set to 37.5 GHz, a typical value for 50G-O/E devices [23]. In all cases, the normalized amplitude $V_{PP,I}/V_{\pi}$ and the pre-compensated length L_C are optimized. For L = 5 km, the optimum L_C is 4 km and 4.5 km for the VNLE and the SQRT + VNLE, respectively. The optimum L_C is 9 km, 14 km, and 18 km for L equal to 10 km, 15 km, and 20 km, respectively. We select the VNLE *NL memory* = 5 samples, 10 samples and 20 samples. The maximum ODN loss gain obtained by increasing the NL memory increases as the fiber length L increases. At very short distance, for example L = 5 km, the improvement is negligible due to the limited nonlinear distortions introduced by the fiber. To achieve 29 dB ODN loss, only *NL memory* = 5 samples is required for $L \leq 14$ km, whereas *NL memory* = 10 samples is required for L in the 15–20 km range.

In PON solutions, the ONU is the most cost-sensitive element. Here we compare the complexity in terms of MPS of different RX DSP options, which are placed at the ONU side in the downstream transmission. The MPS of the SQRT, FFE, and DFE techniques can be calculated as described in [17]. For the VNLE, the MPS can be evaluated by using Equation (5) in [34]. In this paper, the order of VNLE is equal to 3, the linear memory is 121 samples, and the quadratic and cubic memory (NL memory) are equal to 5 samples, 10 samples, 15 samples or 20 samples. The complexity in terms of MPS of the different RX DSP used in the paper is summarized in Table 1. The least complex method that allows us to achieve 29 dB maximum ODN loss with 50G-O/E devices (indicated with an asterisk in Table 1) is the SQRT + VNLE with *NL memory* = 10 samples, requiring 893 MPS. Small increments in NL memory (especially in cubic memory) can introduce a significant increase in complexity.

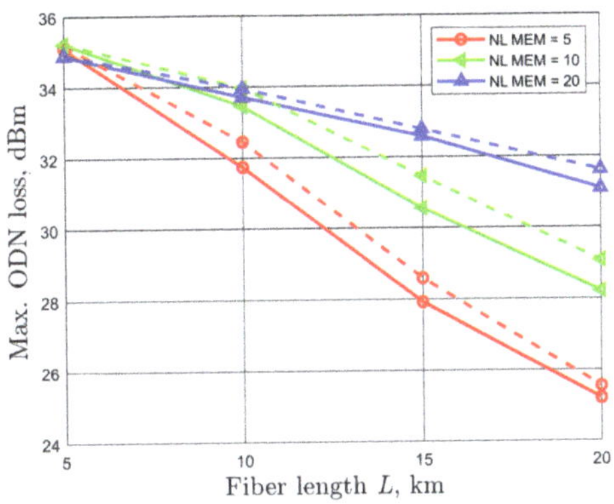

Figure 7. Maximum ODN loss as a function of fiber length targeting $BER_T = 10^{-2}$, for 50G-O/E devices. The normalized amplitude $V_{PP,I}/V_\pi$ and the pre-compensated length L_C are optimized for each case. The transmitted optical power TOP = 15 dBm. Solid: VNLE, Dashed: SQRT + VNLE.

Table 1. Complexity comparison in terms of multiplication per sample (MPS) for different RX DSP options.

RX DSP Options	NL Memory of VNLE	MPS
SQRT [1] + FFE [2] + DFE [3]	-	127
VNLE [4]	5	256
	10	891
	15	2401 *
	20	5161 *
SQRT [1] + VNLE [4]	5	258
	10	893 *
	15	2403 *
	20	5163 *

[1] SQRT requires two additional MPS. It is implemented by using the second order polynomial function as shown in [16]. [2] For FFE, MPS equals the number of FFE taps N_{FFE}. [3] For DFE, MPS equals the number of DFE taps N_{DFE}. Note: DFE requires additional implementation MPS due to the feedback loop. [4] The MPS of VNLE is evaluated by using the method in [34]. * 29 dB maximum ODN loss can be achieved with 50G-O/E devices.

Although VNLE is an effective RX DSP option to compensate for linear and non-linear distortion, its complexity can be extremely high. The SQRT provides a marginal performance improvement but at the cost of a limited complexity enhancement by only 2 MPS. Thus, the SQRT can be used in combination with the VNLE to decrease the NL memory as well as maintain the desired system performance. For example, as shown in Figure 6, to achieve ≥29 dB maximum ODN loss by using 50G O/E, when the VNLE is applied alone, NL memory of 12 samples and 13 samples is required for TX and RX bandwidth of 40 GHz and 35 GHz, respectively, which corresponds to MPS of 1369 and 1668. For the SQRT + VNLE technique, NL memory of 9 samples and 11 samples is needed for TX and RX bandwidth of 40 GHz and 35 GHz, respectively, which corresponds to MPS of 706 and 1111, respectively.

4. Discussion and Conclusions

We demonstrated through simulations the feasibility of 200 Gbps per λ PAM-8 downstream PON transmission in C-band over up to 20 km SMF, preserving DD scheme, achieving 29 dB maximum ODN loss with different RX DSP options. We analyzed the impact of the peak-to-peak amplitude of the signals, which must be optimized to obtain the best system performance. The VNLE and the SQRT can partially compensate for the nonlinear distortions, and allow a higher optimum normalized amplitude $V_{PP,I}/V_{\pi}$, resulting in a higher OMI and better system performance. Using the SQRT + FFE + DFE method, without VNLE, 29 dB maximum ODN loss cannot be reached even considering a very large 70 GHz TX and RX bandwidth. For 20 km transmission with 50G O/E devices, when the NL memory of the VNLE is higher than 12 samples, 29 dB maximum ODN loss can be reached by using the VLNE alone. The minimum required NL memory is reduced to 9 samples when the SQRT is combined with the VNLE. Although the reduction in terms of NL memory is relatively small, the resulting reduction in RX DSP complexity (evaluated in terms of MPS) is remarkable. The SQRT + VNLE requires 706 MPS, whereas 1369 MPS are needed for the VNLE alone. This confirms that the SQRT can be used in combination with VNLE to decrease the NL memory (and eventually the RX DSP complexity) while preserving the required link loss budget. We also studied the impact of different RX DSP options for different fiber length. Our findings show that, 200 Gbps PAM-8 communication can be achieved with 29 dB maximum ODN loss by using a variable number of NL memory equal to 5 samples and 10 samples in the 5–14 km and 15–20 km link length ranges, respectively.

Author Contributions: Conceptualization, R.G., V.F., R.M., G.R., P.T.-F. and H.W.; methodology, R.G., V.F., R.M., G.R., P.T.-F. and H.W.; software, P.T.-F. and H.W.; validation, R.G., V.F., R.M., G.R., P.T.-F. and H.W.; formal analysis, R.G., V.F., R.M., G.R., P.T.-F. and H.W.; investigation, R.G., V.F., R.M., G.R., P.T.-F. and H.W.; resources, R.G., V.F., R.M., G.R., P.T.-F. and H.W.; data curation, H.W.; writing—original draft preparation, H.W.; writing—review and editing, R.G., V.F., R.M., G.R., P.T.-F. and H.W.; visualization, R.G., V.F., R.M., G.R., P.T.-F. and H.W.; supervision, R.G., R.M. and V.F.; project administration, R.G., R.M. and V.F. All authors have read and agreed to the published version of the manuscript.

Funding: This research received no external funding.

Institutional Review Board Statement: Not applicable.

Informed Consent Statement: Not applicable.

Data Availability Statement: Not applicable.

Acknowledgments: We thank the support of the PhotoNext initiative at Politecnico di Torino. This work was carried out in the framework of a research contract with Telecom Italia Lab (the TIM company research laboratory).

Conflicts of Interest: The authors declare no conflict of interest.

References

1. Higher Speed Passive Optical Networks, ITU-T G.9804.x Series of Recommendations. G.9804.1 Consented. July 2019. Available online: https://www.itu.int/rec/T-REC-G.9804.1-201911-I/en (accessed on 25 February 2022).
2. Wey, J.S. The Outlook for PON Standardization: A Tutorial. *J. Lightwave Technol.* **2020**, *38*, 31–42. [CrossRef]
3. Houtsma, V.; Mahadevan, A.; Kaneda, N.; van Veen, D. Transceiver technologies for passive optical networks: Past, present, and future [Invited Tutorial]. *J. Opt. Commun. Netw.* **2021**, *13*, A44–A55. [CrossRef]
4. Zhang, D.; Liu, D.; Wu, X.; Nesset, D. Progress of ITU-T higher speed passive optical network (50G-PON) standardization. *J. Opt. Commun. Netw.* **2020**, *12*, D99–D108. [CrossRef]
5. Wey, J.S.; Zhang, J. Passive Optical Networks for 5G Transport: Technology and Standards. *J. Lightwave Technol.* **2019**, *37*, 2830–2837. [CrossRef]
6. Zou, J.; Sasu, S.A.; Lawin, M.; Dochhan, A.; Elbers, J.-P.; Eiselt, M. Advanced optical access technologies for next-generation (5G) mobile networks [Invited]. *J. Opt. Commun. Netw.* **2020**, *12*, D86–D98. [CrossRef]
7. Tomkos, I.; Klonidis, D.; Pikasis, E.; Theodoridis, S. Toward the 6G Network Era: Opportunities and Challenges. *IT Prof.* **2020**, *22*, 34–38. [CrossRef]
8. De Alwis, C.; Kalla, A.; Pham, Q.V.; Kumar, P.; Dev, K.; Hwang, W.J.; Liyanage, M. Survey on 6G Frontiers: Trends, Applications, Requirements, Technologies and Future Research. *IEEE Open J. Commun. Soc.* **2021**, *2*, 836–886. [CrossRef]
9. van Veen, D.; Houtsma, V. Strategies for economical next-generation 50G and 100G passive optical networks [Invited]. *J. Opt. Commun. Netw.* **2020**, *12*, A95–A103. [CrossRef]

10. Zhang, J.; Yu, J.; Wey, J.S.; Li, X.; Zhao, L.; Wang, K.; Kong, M.; Zhou, W.; Xiao, J.; Xin, X.; et al. SOA Pre-Amplified 100 Gb/s/λ PAM-4 TDM-PON Downstream Transmission Using 10 Gbps O-Band Transmitters. *J. Lightwave Technol.* **2020**, *38*, 185–193. [CrossRef]
11. Saber, M.G.; Gutiérrez-Castrejón, R.; Alam, M.S.; Xing, Z.; El-Fiky, E.; Xu, L.; Cavaliere, F.; Vall-Llosera, G.; Lessard, S.; Plant, D.V. 100 Gb/s/λ Duo-Binary PAM-4 Transmission Using 25G Components Achieving 50 km Reach. *IEEE Photonics Technol. Lett.* **2020**, *32*, 138–141. [CrossRef]
12. Erkılınç, M.S.; Emmerich, R.; Habel, K.; Jungnickel, V.; Schmidt-Langhorst, C.; Schubert, C.; Freund, R. PON transceiver technologies for ≥50 Gbits/s per λ: Alamouti coding and heterodyne detection. *J. Opt. Commun. Netw.* **2020**, *12*, A162–A170. [CrossRef]
13. Che, D.; Iannone, P.; Raybon, G.; Matsui, Y. 200 Gb/s Bi-Directional TDM-PON with 29-dB Power Budget. In Proceedings of the 2021 European Conference on Optical Communication (ECOC), Bordeaux, France, 13–16 September 2021; pp. 1–3. [CrossRef]
14. Torres-Ferrera, P.; Wang, H.; Ferrero, V.; Gaudino, R. 100 Gbps/λ PON downstream O- and C-band alternatives using direct-detection and linear-impairment equalization. *J. Opt. Commun. Netw.* **2021**, *13*, A111–A123. [CrossRef]
15. Torres-Ferrera, P.; Rizzelli, G.; Ferrero, V.; Gaudino, R. 100+ Gbps/λ 50 km C-Band Downstream PON Using CD Digital Pre-Compensation and Direct-Detection ONU Receiver. *J. Lightwave Technol.* **2020**, *38*, 6807–6816. [CrossRef]
16. Wang, H.; Torres-Ferrera, P.; Rizzelli, G.; Ferrero, V.; Gaudino, R. 100 Gbps/λ C-Band CD Digital Pre-Compensated and Direct-Detection Links with Simple Non-Linear Compensation. *IEEE Photonics J.* **2021**, *13*, 7200908. [CrossRef]
17. Torres-Ferrera, P.; Rizzelli, G.; Wang, H.; Ferrero, V.; Gaudino, R. Experimental Demonstration of 100 Gbps/λ C-Band Direct-Detection Downstream PON Using Non-Linear and CD Compensation with 29 dB+ OPL Over 0 Km–100 Km. *J. Lightwave Technol.* **2020**, *40*, 547–556. [CrossRef]
18. Zhu, Y.; Yi, L.; Yang, B.; Huang, X.; Wey, J.S.; Ma, Z.; Hu, W. Comparative study of cost-effective coherent and direct detection schemes for 100 Gb/s/λ PON. *J. Opt. Commun. Netw.* **2020**, *12*, D36–D47. [CrossRef]
19. Suzuki, N.; Miura, H.; Matsuda, K.; Matsumoto, R.; Motoshima, K. 100 Gb/s to 1 Tb/s Based Coherent Passive Optical Network Technology. *J. Lightwave Technol.* **2018**, *36*, 1485–1491. [CrossRef]
20. Zhang, J.; Jia, Z.; Xu, M.; Zhang, H.; Campos, L.A. Efficient preamble design and digital signal processing in upstream burst-mode detection of 100G TDM coherent-PON. *J. Opt. Commun. Netw.* **2021**, *13*, A135–A143. [CrossRef]
21. Zhang, Z.; Guo, Q.; Ju, C.; Cai, S.; Wang, L.; Zhang, M.; Chen, X. Optical- and Electrical-Domain Compensation Techniques for Next-Generation Passive Optical Networks. *IEEE Commun. Mag.* **2019**, *57*, 144–150. [CrossRef]
22. *IEEE Std 802.3ca-2020*; IEEE Standard for Ethernet Amendment 9: Physical Layer Specifications and Management Parameters for 25 Gb/s and 50 Gb/s Passive Optical Networks. IEEE: Piscataway, NJ, USA, 2020; pp. 1–267.
23. Harstead, E.; Bonk, R.; Walklin, S.; van Veen, D.; Houtsma, V.; Kaneda, N.; Mahadevan, A.; Borkowski, R. From 25 Gb/s to 50 Gb/s TDM PON: Transceiver architectures, their performance, standardization aspects, and cost modeling. *J. Opt. Commun. Netw.* **2020**, *12*, D17–D26. [CrossRef]
24. Wey, J.S.; Nesset, D.; Valvo, M.; Grobe, K.; Roberts, H.; Luo, Y.; Smith, J. Physical layer aspects of NG-PON2 standards—Part 1: Optical link design. *J. Opt. Commun. Netw.* **2016**, *8*, 33–42. [CrossRef]
25. Rosales, R.; Cano, I.; Nesset, D.; Brenot, R.; Dubrovina, N.; Durán-Valdeiglesias, E.; Debregeas, H. Achieving high budget classes in the downstream link of 50G-PON. *J. Opt. Commun. Netw.* **2021**, *13*, D13–D21. [CrossRef]
26. Spiekman, L.; Piehler, D.; Iannone, P.; Reichmann, K.; Lee, H. Semiconductor Optical Amplifiers for FTTx. In Proceedings of the 2007 9th International Conference on Transparent Optical Networks, Rome, Italy, 1–5 July 2007; pp. 48–50. [CrossRef]
27. Rosales, R.; Cano, I.N.; Nesset, D.; Brenot, R.; Talli, G. 50G-PON Downstream Link up to 40 km with a 1342 nm Integrated EML + SOA. *IEEE Photonics Technol. Lett.* **2022**, *34*, 306–308. [CrossRef]
28. Characteristics of a Single-Mode Optical Fiber and Cable. ITU-T Recommendation G.652. 2009. Available online: https://www.itu.int/itu-t/recommendations/rec.aspx?rec=13076 (accessed on 25 February 2022).
29. Fritzsche, D.; Lischka, L.; Breuer, D.; Schäffer, C.G. Volterra based nonlinear equalizer with reduced complexity. *Asia-Pac. Opt. Commun.* **2007**, *6783*, 448–455. [CrossRef]
30. Prat, J.; Napoli, A.; Gene, J.M.; Omella, M.; Poggiolini, P.; Curri, V. Square root strategy: A novel method to linearize an optical communication system with electronic equalizers. In Proceedings of the 2005 31st European Conference on Optical Communication, Glasgow, UK, 25–29 September 2005; Volume 3, pp. 713–714. [CrossRef]
31. Poggiolini, P.; Bosco, G.; Carena, A.; Curri, V.; Jiang, Y.; Forghieri, F. The GN-Model of Fiber Non-Linear Propagation and its Applications. *J. Lightwave Technol.* **2014**, *32*, 694–721. [CrossRef]
32. M2 Optics, Inc. Defining Optical Modulation Index. Available online: https://www.globalspec.com/m2optics/ref/Defining_OMI.pdf (accessed on 5 January 2022).
33. Chen, Z.; Yan, L.; Pan, W.; Luo, B.; Zou, X.; Jiang, H. A Transmission Model of Analog Signals in Photonic Links. *IEEE Photonics J.* **2014**, *6*, 7903813. [CrossRef]
34. Tsimbinos, J.; Lever, K.V. The computational complexity of nonlinear compensators based on the Volterra inverse. In Proceedings of the 8th Workshop on Statistical Signal and Array Processing, Corfu, Greece, 24–26 June 1996; pp. 387–390. [CrossRef]

applied sciences

MDPI

Article

Experimental Demonstration and Performance Enhancement of 5G NR Multiband Radio over Fiber System Using Optimized Digital Predistortion

Muhammad Usman Hadi [1,*], Muhammad Awais [2], Mohsin Raza [2], Muhammad Ikram Ashraf [3] and Jian Song [4]

1 School of Engineering, Ulster University, Newtownabbey BT37 0QB, UK
2 Faculty of Computer Engineering, Edge Hill University, Ormskirk L39 4QP, UK; m.awais@edgehill.ac.uk (M.A.); Razam@edgehill.ac.uk (M.R.)
3 Nokia Bell Labs, FI-02610 Espoo, Finland; ikram.ashraf@nokia-bell-labs.com
4 Department of Electronic Systems, Aalborg University, 9220 Aalborg, Denmark; jians@es.aau.dk
* Correspondence: usmanhadi@ieee.org

Abstract: This paper presents an experimental realization of multiband 5G new radio (NR) optical front haul (OFH) based radio over fiber (RoF) system using digital predistortion (DPD). A novel magnitude-selective affine (MSA) based DPD method is proposed for the complexity reduction and performance enhancement of RoF link followed by its comparison with the canonical piece wise linearization (CPWL), decomposed vector rotation method (DVR) and generalized memory polynomial (GMP) methods. Similarly, a detailed study is shown followed by the implementation proposal of novel neural network (NN) for DPD followed by its comparison with MSA, CPWL, DVR and GMP methods. In the experimental testbed, 5G NR standard at 20 GHz with 50 MHz bandwidth and flexible-waveform signal at 3 GHz with 20 MHz bandwidth is used to cover enhanced mobile broad band and small cells scenarios. A dual drive Mach Zehnder Modulator having two distinct radio frequency signals modulates a 1310 nm optical carrier using distributed feedback laser for 22 km of standard single mode fiber. The experimental results are presented in terms of adjacent channel power ratio (ACPR), error vector magnitude (EVM), number of estimated coefficients and multiplications. The study aims to identify those novel methods such as MSA DPD are a good candidate to deploy in real time scenarios for DPD in comparison to NN based DPD which have a slightly better performance as compared to the proposed MSA method but has a higher complexity levels. Both, proposed methods, MSA and NN are meeting the 3GPP Release 17 requirements.

Keywords: digital predistortion; magnitude selective affine; radio over fiber; neural network; error vector magnitude; adjacent channel power ratio

Citation: Hadi, M.U.; Awais, M.; Raza, M.; Ashraf, M.I.; Song, J. Experimental Demonstration and Performance Enhancement of 5G NR Multiband Radio over Fiber System Using Optimized Digital Predistortion. *Appl. Sci.* **2021**, *11*, 11624. http://doi.org/10.3390/app112411624

Academic Editor: Giuseppe Rizzelli

Received: 4 November 2021
Accepted: 1 December 2021
Published: 7 December 2021

1. Introduction

With recent advances in 5G and beyond, the accelerating growth in base stations (BS) has led to centralization of radio access network (RAN) [1,2], which decreases the capital expenditure as it leads to simplifications in network management [2]. To facilitate C-RAN, a fronthaul (FH) connects base band units (BBU) to remote radio heads (RRH) (see Figure 1). With 5G in roll out stage in most part of the developed world, the microwave photonics-based solutions such as Radio over Fiber (RoF) have a higher significance connecting the BBUs with RRUs [3–5] due to advantages such as cost effectiveness; immunity to electromagnetic disturbance, broader bandwidth and increasing the wireless links reach for all type of distances ranging from short to long.

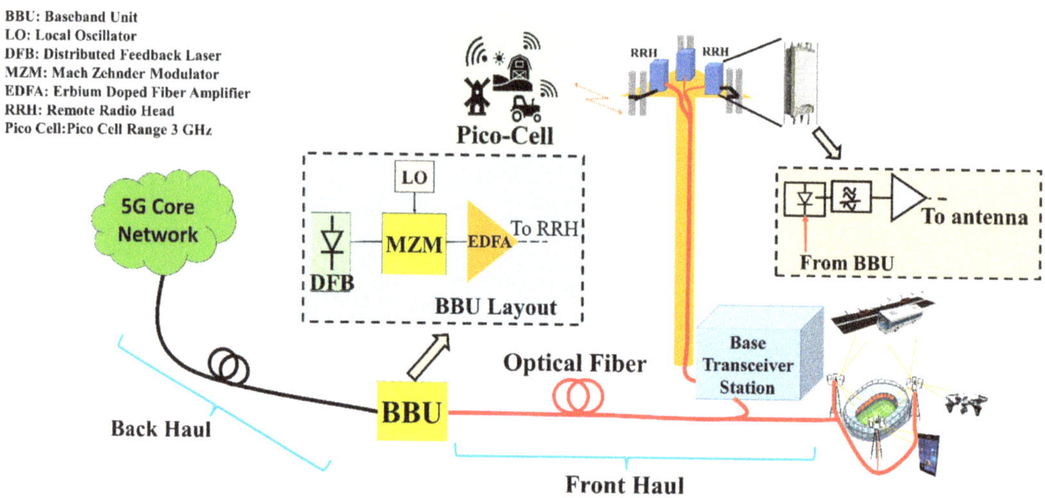

Figure 1. Block diagram of C-RAN based 5G application system comprising of Backhaul, Fronthaul and application scenarios such as stadiums, railways, transport, factories, etc.

There have been various versions of RoF such as Analog Radio over Fiber (A-RoF) [6,7], Digital Radio over Fiber (D-RoF) [8,9], Sigma Delta Radio over Fiber (SD-RoF) [10–12] and other variants that have been proposed recently (see Figure 2). Up to an extent, A-RoF links are the simplest, uncomplicated and economical solution, however, they suffer from nonlinearities arising due to signal impairments and devices involved such as laser modules, fibers and photodiodes. The other solutions comprise of utilizing D-RoF or SD-RoF. Considering D-RoF systems, the requirements of conversion from analog to digital (ADC) and digital to analog converters (DAC) makes the process very expensive. In addition, due to high data rate capacity and requirement of high bandwidth, common public radio interface (CPRI) restrictions are faced. CPRI bottleneck can be overcome by exploiting the SD-RoF. Here, ADCs and DACs are not required as it is based on the sigma-delta modulation which utilizes one bit of ADC but the method is complex hence not preferred. Moreover, the quantization noise is high for 1 bit that requires an additional band pass filter (BPF) at the RRH. However, this additional complexity is not alone to be handled for S-DRoF implementation, the addition of BPF results in additional amplitude and phase noise that requires additional solution for removal of these additional nonlinearities [13].

From this it is evident that exploiting other schemes (D-RoF/SD-RoF) is cumbersome. Hence, comparatively, owing to the legacy, infrastructure, and cost effectiveness of the A-RoF systems, they are the better choice for optical fronthaul (OFH). Now then, our best alternative is to counter the nonlinearities of the A-RoF system. Given that we can do this in a simple and practical manner, the RoF technique proves to be extremely advantageous.

Mitigating the nonlinearities of RoF transmission is essential to utilize the system to its best potential and has become an important subject. Within all these different domains, to counter the prevalent issues, a lot of techniques have been exploited. Amongst them, the one that have been utilized extensively are discussed in Section 2 under literature review.

The contribution of nonlinearities from the laser and to some extent, the photodiode part is important as transmission quality decreases and also, the interference with channels nearby is triggered. However, while considering the long-range networks, the nonlinearities due to the combination of fiber chromatic dispersion and laser frequency chirp are usually the main cause of signal impairment [7]. The Orthogonal Frequency Division Modulated (OFDM) signals, such as the emphasized fifth generation (5G) signal, is also liable to these distortions due to high peak-to average power ratio (PAPR).

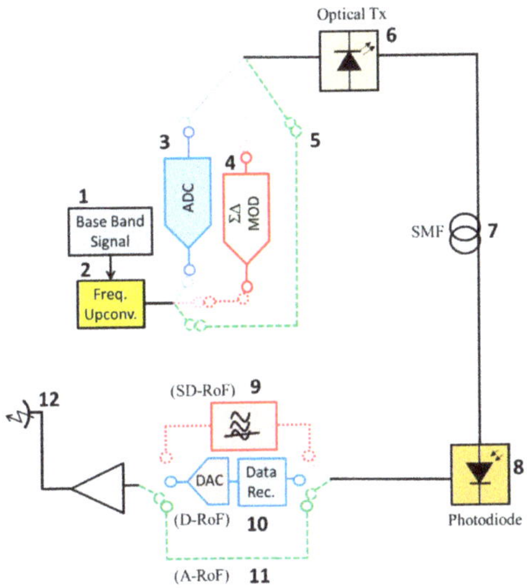

Figure 2. Block diagram of A-RoF, D-RoF and ΣΔ-RoF downlinks. Blocks active for A-RoF: 1-2-5-6-7-8-11-12. Blocks active for D-RoF: 1-2-3-6-7-8-10-12. Blocks active for ΣΔ-RoF: 1-2-4-6-7-8-9-12. BB: Base Band, Freq. Upconv.: Frequency Upconverter, Optical Tx: Transmitter, ΣΔMOD: Sigma-Delta Modulator, Data Rec.: Data Recovery block.

To the best of the authors knowledge, this paper introduces the nonlinearity and signal impairment compensation for OFH based RoF systems utilizing 5G NR based RoF systems. Following a very detailed literature review on the nonlinearities mitigation in Section 2, the novelties of this article are manifold:

1. Firstly, a multiband 5G NR signals are employed in the experimental testbed to cover enhanced mobile broad band (eMBB) scenarios and small cells for 3 GHz and 20 GHz, respectively.
2. A robust DPD technique utilizing negative feedback iteration is shown. The proposed DPD identification method has a relative lower computational complexity as compared toother learning architectures. In the proposed method of DPD identification, firstly, a DPD signal will be identified followed by the estimation of the DPD parameters.
3. The linearization performed is not limited to our previous proposed methods coming from volterra series but includes the "out of box" approach that includes CPWL and DVR method. In addition, a novel magnitude selective affine (MSA) method is proposed that reduces the overheads of complexity such as multiplications in CPWL architecture, however achieving similar efficiency as of CPWL.
4. In addition, a simple optimized neural network (ONN) based DPD algorithm is proposed that is an upgradation of our previously proposed DPD based method utilizing deep learning to perform linearization of 50 MHz 5G New Radio (NR) based RoF links. The NN DPD method is executed by a different type of training which does not utilize Indirect Learning Architecture (ILA). Initially we emulate the generic RoF link using a RoF NN and then train the proposed DPD ONN using this, by backpropagating the errors.

For the first time, a comparative experimental study has been held where MP, GMP, DVR, CPWL, MSA and ONN are compared for 5G NR multiband signal. The performance is evaluated in terms of Adjacent Channel Power Ration (ACPR) Error Vector Magnitude (EVM) and complexity in terms of multiplications and coefficients requirements. The summary of this work is shown in the Figure 3 where overall summary of each respective section is shown.

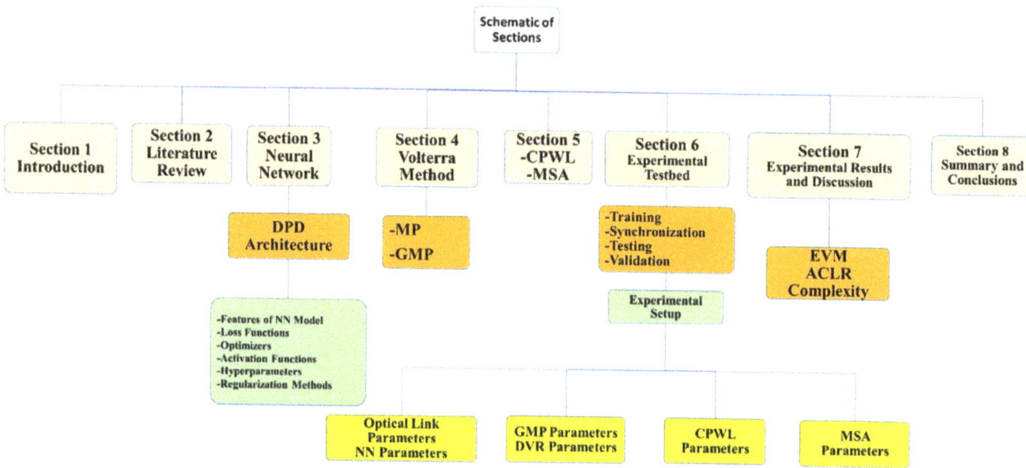

Figure 3. Organization of our work in this paper (Sections 1–8).

2. Literature Review

In this section, the linearization methodologies that have been implied for the RoF systems are discussed. The linearization methodologies consist of Electrical, Optical and Machine Learning Methods. Electrical methods are subdivided in to analog and digital methods. Digital methods are subdivided into predistortion and postdistortion methods. Optical methods consist of largely dual wavelength, singular polarization, mixed polarization, etc., while a newer avenue of linearization has been coined that belongs to machine learning. The higher schematic of these methods is summarized in the Figure 4.

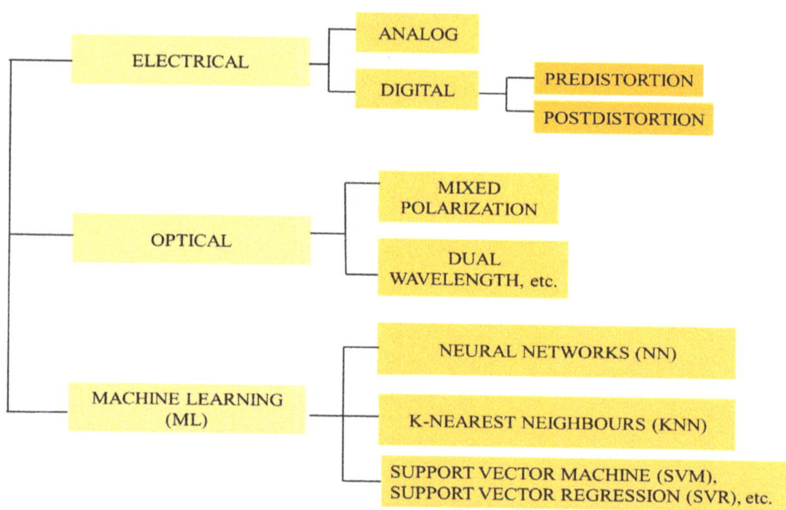

Figure 4. Different types of linearization.

A detailed literature review is enlisted for the mitigation of the impairments of RoF system that are categorized in Table 1. The table summarizes the method employed, type of linearization, category, parameters evaluated, advantages and disadvantages of respective methods.

Table 1. Methods exploited to mitigate nonlinearities.

No.	Author	Type	Subcategory	Parameter	Linearization	Pros	Disadvantages
1	Draa et al.; Chen et al. [14,15]	Electrical	Analog Predistortion	IMD3	Complete RoF system (Laser, photodiode, LNA)	Phase maintenance for IMD3 generated components	Perplexing suppression of second order nonlinear distortion for large bandwidth
2	Hass et al. [16]	Optical	Mixed Polarization	Second/third order nonlinear distortion	Complete RoF system	Suppression of second and third order nonlinearities	RoF transmission must be polarization dependent. Compression of linear components to some extent.
3	Zhu et al. [17]	Optical	Dual wavelength linearization (DWL)	Second/third order nonlinear distortion	Complete RoF system	Suppression of second and third order nonlinearities	Wavelength dependent transmission, i.e., suppression of nonlinear components for anti-phased wavelengths only.
4	Ghannouchi et al. [18]	Digital	Digital Predistortion (DPD)	Third order nonlinearities	Power Amplifier	Wideband linearization achievable	DSP required is complicated. Consumption of power is huge.
5	Duan et al. [19]	Digital	DPD	ACPR, EVM	Laser	Added accuracy with less DSP requirements	High complexity and calculation time for digital linearization for higher nonlinearity order and memory depth.
6	Pei et al. [20]	Digital	DPD	ACPR	Laser/RoF	Higher suppression in ACPR by 15 dB	Feedback complexity.
7	Lam et al. [21]	Digital	Digital Post Processing	ACPR, BER	RoF	All order nonlinear distortion components significantly compressed.	High speed digitizer required. Only applicable to uplinks. Deployment of DPP at RRH side, so prototype price passes to customer side, adding complexity to the RRH.
8	Hekkala et al. [22]	Digital	DPD	ACPR, EVM	Laser only	Less complexity and over head	EVM improved by 0.3%. Link length not considered. Intermediate frequency used for DPD. RoF link was not composed of laserfiber-photodiode only. Attenuators and amplifiers were used so perhaps the signal impairments were corrected due to these components. Limited to sinusoid I/P signal.
9	Hadi et al. [23]	Digital	DPD	C/HD2, IIP2, IIP3	Combination of fiber dispersion and laser chirp	Linearizes links up to tens of km	Limited to sinusoidal (single/dual) I/P tones. No ACPR, EVM shown.
10	Vieira et al. [24]	Digital	DPD	EVM	Laser	Utilization of OFDM signal with 5 MHz bandwidth	Only magnitude (AM/AM) linearization shown. RoF link is not generic, contains 10 dB attenuator.
11	Hekkala et al. [25]	Digital	DPD	ACPR, BER	Laser	Utilization of OFDM signal with 12.5 MHz bandwidth	Joint compensation proposed for PA and RoF hence reducing improvement.
12	Mateo et al. [26]	Digital	DPD	EVM, ACPR	RoF system	Utilization of LTE 20 MHz signal	RoF link is not generic; contains signal amplification by LNA and a PA. Unrealistic feedback for 10 km length; requires an uplink for the linearization of downlink.
13	Mateo et al. [27]	Digital	DPD	NMSE, ACPR	RoF system	Utilization of LTE 20 MHz with 16 QAM modulation	PSD of O/P with and without DPD is different after normalization, DPD reduced the bandwidth of the signal. DFB laser has not been pushed to higher RF I/P powers to see the efficacy of the predistorter.
14	Mateo et al. [28]	Digital	DPD	ACPR, EVM	RoF system	Linearization considering ideal and no feedback	Results are attenuation dependent, i.e., results similar to ideal case can be obtained with proper attenuation using different optimization algorithms.
15	Roselli et al. [29]	Electrical	Analog Predistortion	IMD3	Laser	Maintenance of phase for generated IMD3 components	Large scale production difficult as variant predistorter required per RoF transmitter.

Table 1. *Cont.*

No.	Author	Type	Subcategory	Parameter	Linearization	Pros	Disadvantages
16	R. B. Childs et al. [30]	Electrical	Analog Predistortion	IMD3	Laser	Accurate correction	Tough synthesis of arbitrary transfer functions in the time domain; due to coupling between different paths, simultaneous correction of both, second and third order is not possible.
17	Veiga et al. [31]	Electrical	Analog Predistortion	IMD3	Laser	Phase maintenance is easy	To compensate arbitrary bandwidth limitation in the frequency domain.
18	Hadi et al. [32]	Digital	Direct DPD	EVM, ACPR, IMD	Laser	Only requires transient chirp coefficient, no exhaustive training.	Limited to laser; works only for kilometers of length; complex behavior.
19	Hadi et al. [33]	Digital	DPD for short link	NMSE, ACPR	Laser	Feasible closed loop DPD	Training is time consuming and complex
20	Hadi et al. [34]	Digital	DFB based	NMSE, ACPR, EVM	Laser	Utilization of DVR, GMP, MP	High complexity. I/Q imbalance.
21	Hadi et al. [35]	Digital	ML-NN based	ACPR, EVM	Laser	Learns nonlinearities of model in contrast with Volterra based methods, avoids requirement of explicit designing of distortion causes	Limited to LTE framework. Upgradation to 5G ambiguous.

Along with the numerous techniques listed above, mitigation of the impairments of RoF system have also been studied previously [24–35]. A feedforward scheme has also been analyzed [26,27] but is complex and to counter these problems, optical methods such as dual parallel modulation [28], mixed polarization [29,30], etc. have been presented. In addition to the nonlinearities of optical fiber/modulators, RoF is liable to other issues such as amplified emission noise and phase noise [32,33] that further limit performance of RoF increasing proportionally with RF carrier frequency increase.

Moreover, Digital Signal Processing (DSP) methodologies have been exploited as well owing to their adaptive flexibility. A lot of such techniques have been studied to counter the phase noise issue [36–38], of optical modulator [39,40]. Digital pre-distortion is presented for laser resonance, improving results in [41]. Apart from DSP approaches, Digitized RoF (DRoF) has also been utilized for nonlinearity mitigation [42–45].

Another important concept proposed, Sigma Delta RoF (SD-RoF) is a method involving the plus points of the DRoF in addition to its own implementation [46,47]. This is especially advantageous when utilizing optical transmitters having low linearity. On the other hand, considering Machine Learning as a domain of nonlinearity compensation, a lot of models and methodologies have been proposed such as the KNN/SVM [48]; SVM [49,50]; k-means [51]; Convolution Neural Network (CNN) and Binary-CNN (BCNN) [52,53] and Fully-CNN (FCNN) [54–61].

3. Neural Network Based DPD Architecture

Neural Networks (NN) are sophisticated networks, similar to the suggested NN based DPD model and they require extensive training data. This model is then cascaded with the RoF link, but its output is not known. However, in the case of an RoF link, the output is known hence, we form an RoF NN model and train it to mirror the original RoF link. Once we have formulated this RoF NN, we can now backpropagate through it and update the parameters in the suggested NN DPD model.

Supposing that the considered RoF link has a transfer function $H(n)$ and an output signal $y(n)$; and a baseband signal $x(n)$ must be sent through it, DPD aims to calculate the inverse transfer function of this RoF link denoted by $\hat{I}^{-1}$, whose output will then be denoted by $\hat{x}(n)$.

This can be expressed by:

$$\hat{x}(n) = \hat{I}^{-1}(x(n)) \quad (1)$$

While,

$$y(n) = Gx(n) = I(\hat{x}(n)) \tag{2}$$

where G is the gain. The NN here calculates $\hat{I}^{-1}$ utilized for predistortion. A direct training cannot be implemented for construction of the NN for DPD as the ideal $\hat{x}(n)$ is not known.

The option analyzed for performance assessment is illustrated in Figure 5 [1].

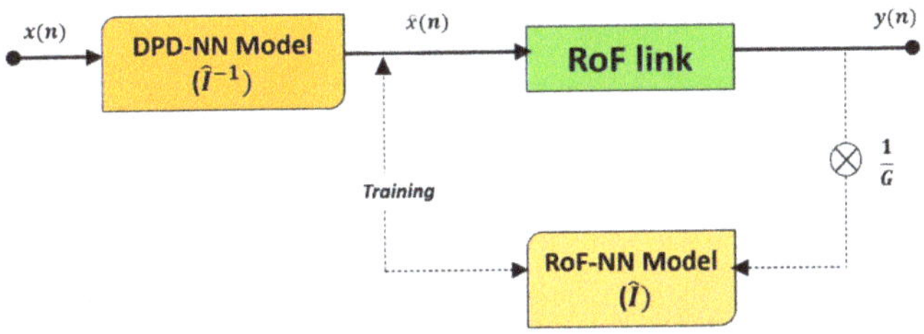

Figure 5. Block diagram of RoF system utilizing RoF NN model and NN based DPD system. The RoF NN model $\hat{I}$ is obtained upon transmission of I/P and O/P through the generic RoF link. Once achieved, we train I^{-1} by backpropagating error through $\hat{I}$. Then, linearization is performed by connecting this DPD-RoF model to generic RoF link. Predistortion is held in a digital baseband eliminating DACs and ADCs.

Initially, the RoF link is emulated by the second NN. The generic RoF link has $\hat{x}(n)$ as input and $\frac{y(n)}{G}$ as output data, considering which, training is executed with the regression based NN model for them. This results in learning of the NN and now an approximate transfer function $\hat{I}$ can be identified by it. Upon the formation of the RoF NN model, the model weights are fixed after which we connect it with the NN DPD model. Now, we use the original input, i.e., $x(n)$ and output as training data to calculate error using a loss function. We then backpropagate it via $\hat{I}$ to train $\hat{I}^{-1}$.

3.1. Features of the NN Model

This section will discuss all the features of the Neural networks that are essential to be used in DPD based RoF system. The summary of the topics discussed in this section is given as follows. Firstly, Section 3.1.1 discusses the loss functions while Section 3.1.2 talks about optimizers. Section 3.1.3 talks about activation functions and Section 3.1.4 talks about hyperparameters. Section 3.1.5 discusses the regularization methods and Section 3.1.6 talks about are characteristics of NN.

The NN are layered structures that have been inspired from the human nervous system and are an adequate choice due to their powerful accuracy at approximating any nonlinear function and learning the relationship between their inputs and outputs. The basic working involves an input signal provided at the input layer, processing which takes place in the hidden layer/layers and the output is produced by the last (output layer). Each layer represents a cluster of neurons which are not connected to one another within the layer but are connected to the neurons of the next layer.

3.1.1. Loss Function

Prior to optimizing, it is essential that, the error is estimated for the current state of the model. For this, selecting an error function, normally known as loss function is necessary as it helps calculate the loss of the model. After this, the weights of the model are updated which minimizes this loss function for further evaluations. Loss functions broadly classify into:

i. Regression Loss.
ii. Classification Loss.

Regression Loss Functions

Regression loss functions are divided into following:

Mean Squared Error (MSE)

MSE calculates the average of the squared differences between actual and predicted value. Outcomes are always positive values irrespective of the sign of the actual and predicted values. The minimum value it can result in is 0.0. Note that, larger differences cause more error due to squaring of the value, penalizing the model for them.

Mean Squared Logarithmic Error (MSLE)

As the name suggests, here, the natural logarithm of each of the predicted values is calculated and then the MSE is calculated. MSLE relaxes the punishing effect of large differences in large, predicted values.

Mean Absolute Error (MAE)

The MAE is more powerful. It calculates the average of the absolute difference between the predicted and actual values.

Classification Loss Functions:

Classification loss functions are divided into following:

Binary Cross-Entropy

This function is used when our target values belong to the set $\{0, 1\}$. A score which summarizes the average difference between the predicted and actual probability distributions is calculated and is then minimized. The least score that can be achieved by this function is 0.

Hinge Loss

This function is used when our target values belong to the set $\{-1, 1\}$. This function basically promotes correctness of the sign of values and will allot more error for differences in signs between predicted values and actual values.

Squared Hinge Loss

This is a function obtained by modifications in the hinge loss function. It merely calculates the square of the hinge loss score. This causes the error function to smoothen making it easy to work with numerically. It smoothens the surface of the error function and makes it easy to work with numerically. As stated in the case of hinge loss, values must belong to the set $\{-1, 1\}$.

Multi-Class Cross-Entropy

Hinge loss is considered for binary classification similarly, multi-class cross entropy loss is considered for multi-class classification. Every class is given a distinct integer value. A score for all classes is summarized by calculating the average difference between the predicted and actual probability distributions. The minimum score that can be obtained is 0.

3.1.2. Optimizers

The process of optimizing loss functions improves the function of the model. The procedure to minimize/maximize a mathematical function is called optimization and utilizes optimizers. Optimizers work by changing the parameters of NNs such as the learning rates or weights. The types of optimizers are:

Gradient Descent (GD)

The GD algorithm works on the principle that the direction opposite to that of the calculated gradient will locate the lower point/surface of the considered function, i.e., the potential minima. It then repeatedly takes steps in that direction with every iteration. In the case of GD, we require large memory as we calculate the gradient over the entire dataset after which the updates are performed meaning that, for huge data, it may take years to converge.

Stochastic Gradient Descent (SGD)

To make up for the problem faced by the GD algorithm, SGD was developed. The SGD algorithm works by computing derivatives by considering only one point at once. This solves the large memory requirement problem; however, time is still an issue as one epoch of SGD takes more time when compared to the GD algorithm.

Mini Batch Stochastic Gradient Descent (MB-SGD)

This algorithm was developed to overcome time complexity problems. MB-SGD as the name suggests, works by considering a small batch of points or a small subset of the whole data at once to compute the derivatives. It takes longer to converge than GD does and additionally, its weight updates are noisy.

SGD with Momentum

SGD with momentum (momentum for short) calculates another parameter, momentum in addition to the gradient in every step. This momentum is the cumulative movement from all previous steps, overcoming the issue of noisy weight updates in MB-SGD by denoising gradients. It moves faster owing to the accumulated momentum which helps it prevent local minima and plateau regions. In addition, the algorithm Nesterov Accelerated Gradient (NAG) is similar to momentum except, it is a future considering method. The cost function is evaluated using future parameters rather than the current ones to prevent the chance of skipping the minima in case of high momentum value.

Adaptive Gradient Descent (AdaGrad)

AdaGrad, works by accumulating the sum of gradient 'squared'. As the name suggests the learning rate is adaptive and doesn't need to be tuned manually. The adaptive learning rate property helps escape typical complexities of non-convex functions such as saddle points as they take a straight path compared to methods such as GD, SGD, MB-SGD and even momentum. There are chances that these methods become stuck at saddle points and never converge to minima. Unfortunately, AdaGrad is very slow as the sum it accumulates continues to grow resulting in the learning rate to become very small, this consequently leads to vanishing of the gradient.

Adaptive Delta (AdaDelta)

As we know, AdaGrad is a slow converging method due to its large value of the accumulated sum of gradient squared. AdaDelta was developed to resolve this by modifying certain aspects of AdaGrad. Here, rather than storing all the past gradients inefficiently, we recursively define the sum of gradients as a decaying average of all the past squared gradients. This allows AdaDelta to continue learning even after multiple updates.

RMSprop (Root Mean Square Propagation)

RMSprop, similar to AdaDelta, was developed to counter AdaGrad's problem of vanishing learning rates. RMSProp fixes the issue by adding a decay factor which emphasizes on recent gradients neglecting the older ones. An exponentially decaying average of squared gradients is used to divide the learning rate. AdaGrad might keep up with RMSProp initially, but the sum of gradients squared for AdaGrad accumulate and become huge, eventually AdaGrad practically stops moving, in contrast with RMSProp, smartly stored the squares due to the decay rate. This makes it faster than AdaGrad and it impedes search in direction of oscillations.

Adaptive Moment Estimation (ADAM)

The Adam algorithm combines the features of both Momentum and RMSProp meaning that, along with accumulating the exponentially decaying average of the previous squared gradients, it also holds exponentially decaying average of previous gradients as well. Adam is an efficient optimizer due to the component of speed and the capability to adapt gradients acquired from momentum and RMSProp, respectively. This is also the reason it is the most used optimizer.

Comparison of Optimizers

There are two metrics to determine the efficacy of an optimizer, speed of convergence to the global minimum and generalization (the model's performance on new data). Performance of the optimizers and hence their choice is also dependent on the type of data provided, based on that, two types of functions exist- convex and non-convex (Figure 4). In case of the latter, optimizers must be carefully chosen as it has multiple hurdles where our algorithm might become stuck, some of which we will discuss ahead. The performance of mentioned optimizers on different obstacles faced on non-convex functions are discussed below (see Table 2).

Table 2. Comparison of Optimizers.

Obstacle Type	Brief Definition	Convergence Speed (Descending Order of Converging Speed)
Saddle point	Point on the surface of a function which is not a local extremum of the function but the slopes in orthogonal directions are all zero.	ADAM, RMSprop, AdaDelta, AdaGrad, Momentum.
Plateau region	Plateau of a function is a part of its domain where the function has constant value.	ADAM, Momentum
Local minima	Minimum of a function other than the global minimum (extremum).	ADAM, RMSprop, Momentum, AdaGrad

From the table, we see that ADAM has performed comparatively better than others, especially in terms of convergence speed which is why it is the most used and most efficient optimizer.

3.1.3. Activation Functions

The activation or capability of response of a node is not pre-defined. This is deduced with the help of an Activation. It does this by building a relationship between various weights and biases affecting the node and applying the relationship to the node as a function, hence determining responses. It also helps it learn the complex patterns of data. They transform the incoming signals in the node into an output signal which will be used in the next layer of the network or will be the output. The different types of activation functions are summarized in the table below (Table 3).

Table 3. Activation Functions.

Activation Function	Mathematical Representation	Brief Description
Binary Step	$(x) = \begin{cases} 0 \;; for\ x < 0 \\ 1 \;; for\ x \geq 0 \end{cases}$	only has values 1 or 0 as output.
Linear	$y(x) = x$	Linear over entire range $(-\infty, \infty)$.
Sigmoid/Logistic	$y(x) = \frac{1}{1+e^{-x}}$	Range (0, 1)
ReLu (Rectified Linear Unit)	$f(x) = max(0, x)$	Range $[0$ to $\infty)$
Leaky ReLu	$f(x) = max(0.01 * x\ ,\ x)$	Range $(-\infty, \infty)$.
Softmax	$softmax(z_i) = \frac{\exp(z_i)}{\Sigma_j \exp(z_i)}$	Range (0, 1) and sum of O/Ps = 1
tanh/Hyperbolic	$y(x) = \frac{\sinh x}{\cosh x}$	Range $(-1, 1)$

When considering ReLu, the possibility of a vanishing gradient is lower, in contrast with the gradient of sigmoids, which grow smaller with an increase in the absolute value of x.

ReLu has a constant gradient, helping it learn faster making it the better choice, provided we can work in the positive range of input values.

3.1.4. Hyperparameters

The implementation of algorithms on data has certain pre-requisites. There are several hyperparameters that must be defined for the functionality of our network. These are overall functionality defining components and are not to be confused with the internal model parameters. This section discusses the hyperparameters definitions in Table 4.

Table 4. Definition of Hyperparameters.

Hyperparameter	Definition	Explanation
Batch size	Number of samples processed while updating the internal parameters of a model.	GD = Size of Training Set; SGD = 1; B-SGD = 1 < Batch Size < Size of Training Set
Epoch	Number of complete passes through the samples being trained.	For MB-SGD if batch size = 25, then number of complete passes through these 25 samples is number of epochs.
Regularization	Penalizes weight matrices to resolve overfitting and underfitting	Methods-L1, L2, Dropout.

Note that there are no pre-defined fixed values for these hyperparameters. There are some conventionally used values, but they are not mandatory and attaining a value suitable to a function is a trial-and-error method.

3.1.5. Different Methods of Regularization

All datasets are divided into two sections, namely, the training data and the testing data. Sometimes, we come upon conditions where models do not result in good performance on testing data despite performing well on training data. These situations arise due to underfitting and overfitting in Figure 6. Regularization is the procedure in which weight matrices of nodes are penalized, hence overcoming these problems.

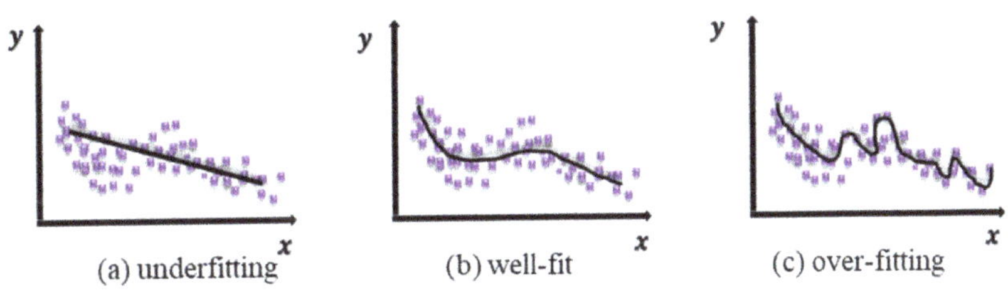

Figure 6. Graphical representation of the training curve (black) learning the training data (purple, scattered) in three labelled cases.

L2 and L1 Regularization

L1 and L2 are simple and widely known types. A regularization term is considered while updating the cost function in this method. This regularization term causes a reduction in the value of the weight matrices as they result in a simpler model, significantly reducing the problem of overfitting.

In L2, we have:

$$Loss = error + \lambda \sum_{i=1}^{N} \left| w_i^2 \right| \quad (3)$$

where λ is the regularization parameter (which is updated for more precise outcomes) and w is weight. The L2 method is referred to as a weight decay procedure because it compels the weights to decay towards zero although never exactly becoming zero.

In L1, we have:

$$Loss = error + \lambda \sum_{i=1}^{N} |w_i| \tag{4}$$

As observed in Equation (4) here, the absolute weight value is considered. Contrary to the L2 method, the weights here can become zero which is beneficial for compression of models.

Dropout

Dropout is the most used type of regularization due to its efficient results. It performs several iterations on the NN and at every iteration, random nodes are selected and removed along with all of their input and output signals/connections as seen in Figure 7. This results in selection of distinct nodes and hence distinct outputs per iteration. More randomness is expressed by splitting one network into its subsets leading to better performance when compared to a single dense model. The method can be implemented on input as well as hidden layers making it an adequate choice for large NNs.

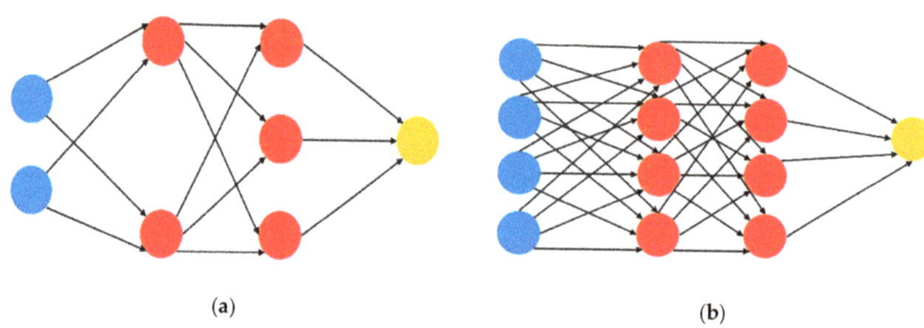

Figure 7. Schematic diagram of complete densely connected in (**a**) NN and (**b**) representation of network upon selection and removal of random nodes, after any one iteration.

3.1.6. Characteristics of the NN Models

Based on the discussion made so far, we have built a foundation of all components of a Neural Network and established that the two neural network models we require here are, first, the DPD NN, used to predistort our original RoF link and second, the emulated RoF NN, which is essential for training this DPD NN. The NN employed here contains *N* hidden-layers, *K* neurons per hidden-layer and is a feedforward fully connected network. The symbolic structure of the employed NN is shown below in Figure 8. The NN has two inputs and two outputs, in view of the complex nature of baseband signals representing both the real and imaginary parts of the signal. The ReLu activation function has been utilized for at least one of the multiple hidden layers (owing to its lower complexity).

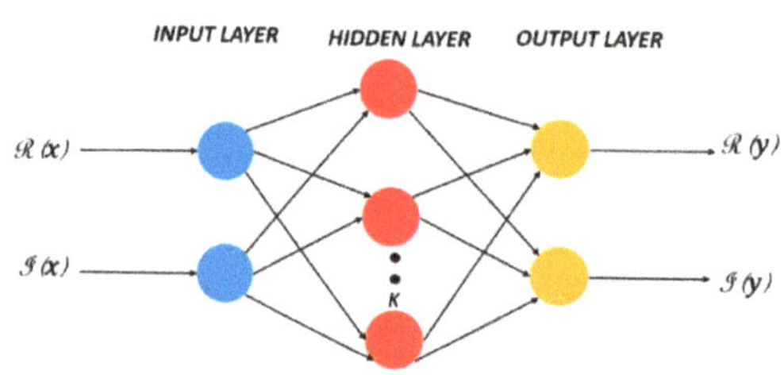

Figure 8. Schematic diagram of the feedforward fully connected NN structure used, with one of the *N* number of hidden layers having *K* neurons per hidden layer. R(*x*) and R(*y*) are real I/P and O/P; I(*x*) and I(*y*) are imaginary I/P and O/P, respectively.

We represent the O/P for the first hidden-layer as follows:

$$l_1(n) = f\left(W_1\begin{bmatrix}\mathbb{R}(x)\\ \mathbb{I}(x)\end{bmatrix} + b_1\right) \quad (5)$$

where, l_1 expresses the first hidden O/P layer, f is the activation function (nonlinear) and W_1 is the weight and b_1 is the bias for the first output layer in the network.

We represent the general output for the ith layer as:

$$l_i(n) = f(W_i h_{i-1}(n) + b_i) \quad (6)$$

where, $i \in \mathbb{N}\ : 2 \leq i \leq N$.

Then, the final output after N hidden layers will be:

$$\hat{x}(n) = W_{N+1}h_N + b_{N+1} \quad (7)$$

Training Algorithm

The algorithm used to train the NN DPD model is given below. The algorithm utilizes MSE as the loss function, the optimizer used is ADAM and backpropagation is used for updating the weights. The process is repeated for a certain number of iterations (Z) to refine performance.

As mentioned earlier, we train our emulated RoF NN model using the I/P and O/P of the original RoF link and once this model is acquired, we connect the NN DPD model to it. Then, upon convergence of this training, we connect the actual RoF link with this DPD NN and proceed with its predistortion (see pseudocode in Algorithm 1).

Algorithm 1. DPD training

$\hat{x}(n) \leftarrow x(n)$
for $i \leq Z$ **do**
$y(n) \leftarrow I(\hat{x}(n))$: //RoF-Transmission
$\hat{I} \leftarrow$ Train on $\hat{x}(n)$, $\frac{y(n)}{G}$ //Updating RoF NN
//Freeze NN weights of $\hat{I}$
$\hat{I} \leftarrow$ Train on $x(n)$. //Use $\hat{I}^{-1}(\hat{I}(x(n)))$
$\hat{x}(n) \leftarrow \hat{I}^{-1}(x(n))$: //Predistort
end for

4. Comparison with Volterra Method

It will be interesting to see the comparison of NN DPD methodology with conventional GMP method that has been validated in [1,6,34]. MP/GMP are the most viable solutions that have been used for DPD.

Volterra series-based models are very commonly used. Upon conversion of signals from electrical to optical domain and vice-versa, certain memory effects are introduced. To take these effects into account, Volterra Series is considered.

We represent Volterra series as:

$$\check{y}(n) = \sum_{k=1}^{+\infty}\int_0^{+\infty}\cdots\int_0^{+\infty} h_k(\tau_1,\ldots,\tau_k)\prod_{r=1}^{k}\check{x}(t-\tau_1)d\tau_1 \quad (8)$$

where $\check{y}(n)$ represents RF out put, $\check{x}(n)$ represents RF input signal and h_m is the mth order Volterra kernel. RF signal is down converted to baseband, and we obtain the envelope through a low pass filter. Then, with input $\check{x}(n)$ baseband image in discrete time is:

$$y(n) = \sum_{k=0}^{k}\sum_{q_1=0}^{Q-1}\sum_{q_2=l_1}^{Q-1}\cdots\sum_{q_{2k+1}}^{Q-1} h_{2k+1}(l_1,\ldots,l_{2k+1})\prod_{r=1}^{k+1}\check{x}(t-l_j)\prod_{r=k+2}^{2k+1}x^*(n-l_j) \quad (9)$$

where, K is the nonlinearity order and Q is the memory depth.

4.1. Memory Polynomial Model (MP)

The Memory polynomial (MP) model is capable of resolving both memory effects and nonlinearities simultaneously. It is also referred to as diagonal Volterra model as all its diagonal terms are non-zero. It is the middle ground between a memoryless model and a full Volterra model due to existence of diagonal memory (non-diagonal terms are zero). The MP model is generally used as it is less complex compared to Volterra series. It can also emulate the nonlinear behavior of Power Amplifiers to an extent which is why it has been used to model Power Amplifiers previously. The equation is as follows [24–35]:

$$y(n) = \sum_{k=0}^{k-1} \sum_{q=0}^{Q-1} C_{kq} x(n-l)|x(n-l)|^k \tag{10}$$

where K is non-linearity order, Q is the memory depth, $y(n)$ is input sequence of the predistorter, $x(n)$ represents baseband input and c_{kq} is the model coefficient.

4.2. Generalized Memory Polynomial (GMP)

GMP has been productively utilized for DPD linearization of Power Amplifiers. Here, we will now use it to aid linearization of RoF links using the Digital predistortion method. If we observe the equation below, we see that, the GMP model, in contrast with MP, holds both, memory of diagonal terms along with crossing terms as well, which is why it outperforms MP and in the coming sections, we will evaluate comparisons with GMP only having established it is superior to MP [1,24–35]:

$$\begin{aligned} y(n) = &\sum_{k=0}^{k_a-1} \sum_{q=0}^{Q_a-1} c_{kq} x(n-q)|x(n-q)|^k \\ &+ \sum_{k=1}^{k_b} \sum_{q=0}^{Q_b-1} \sum_{r=1}^{R_b} d_{kqr}\, x(n-q)|x(n-q-r)|^k \\ &+ \sum_{k=1}^{k_c} \sum_{q=0}^{Q_c-1} \sum_{r=1}^{R_c} e_{kqr}\, x(n-q)|x(n-q+r)|^k \end{aligned} \tag{11}$$

where $x(n)$ is the DPD input and $y(n)$ is the DPD output.

The complex coefficients c_{kq}, d_{kqr} and e_{kqr} denote signal and the envelope; signal and lagging envelope and signal and leading envelope, respectively. K_a, K_b, K_c are the maximum nonlinearity orders, Q_a, Q_b, Q_c are the memory depths. q, r represents the indices of the memory k is nonlinearity index and R_c and R_b exhibit the leading and lagging delay tap lengths, respectively. Since, in [62–71], it was established that GMP is better as compared to MP, therefore for simplicity, evaluation with GMP is included in this paper.

5. Models Based on Canonical Piece Wise Linearization

The DPD linearization has been achieved with Volterra based methods as discussed in Section 3. However, "out of the box" approach which can achieve better performance is always interesting addition to the topic. Recently, it was shown in [26–28] that out of all the possible architectures, CPWL method outperforms the other models such as memory polynomial (MP) and generalized memory polynomial (GMP). CPWL is an obvious choice due to the performance enhancement that it brings, however it has a lot of complexity and overheads. The CPWL model can be expressed as [28]:

$$\begin{aligned} y(n) &= \textstyle\sum_{m=0}^{M}\sum_{k=0}^{K}\sum_{l=1}^{L} c_{m,k,l}^{(1)}\Big||x(n-k)|^2-\beta_l\Big|x(n-m-k) \\ &+\textstyle\sum_{m=1}^{M}\sum_{k=0}^{K}\sum_{l=1}^{L} c_{m,k,l}^{(2)}\Big||x(n-k)|^2-\beta_l\Big|x^2(n-k)x^*(n-m-k) \\ &+\textstyle\sum_{m=1}^{M}\sum_{k=0}^{K}\sum_{l=1}^{L} c_{m,k,l}^{(3)}\Big||x(n-k)|^2-\beta_l\Big|x(n-k)\,|x(n-m-k)|^2 \\ &+\textstyle\sum_{m=1}^{M}\sum_{k=0}^{K}\sum_{l=1}^{L} c_{m,k,l}^{(4)}\Big||x(n-k)|^2-\beta_l\Big|x^*(n-k)\,x^2(n-m-k) \end{aligned} \tag{12}$$

Here, the baseband input is represented by $x(n)$ and the output baseband signal is represented by $y(n)$, K is the FIR length filter, M is represented by memory depth, L is the number of partitions in the CPWL, β_l shows the threshold while $c_{m,k,l}^{(1)}, c_{m,k,l}^{(2)}, c_{m,k,l}^{(3)}, c_{m,k,l}^{(4)}$ presents the model coefficients.

In the Equation (1), there are many orders of multiplications and additions which will add a lot of overhead in terms of complexity and utilization if hardware resources during DPD implementation, the most important of them is dedicated hardware adders and multipliers.

5.1. Decomposed Vector Rotation (DVR) Model

It is a derived from the canonical piecewise-linear (CPWL) function and is a modified form of it. The model's nonlinear function is constructed from piecewise vector decomposition and is entirely different to the one used in previously discussed Volterra series. The model is more flexible and can perform better even with a comparatively small coefficient number than the conventional models. DVR is expressed as:

$$\begin{aligned} y(n) = &\sum_{i=0}^{Q_{DVR}} a_i|x(n-i)| + \sum_{k=1}^{K_{DVR}}\sum_{i=0}^{Q_{DVR}} a_{ki,1}\,||x(n-i)-\beta_k|e^{j\theta(n-1)} \\ &+\sum_{k=1}^{K_{DVR}}\sum_{i=0}^{Q_{DVR}} a_{ki,21}\,||x(n-i)-\beta_k|e^{j\theta(n-1)}.|x(n)| + \sum_{k=1}^{K_{DVR}}\sum_{i=0}^{Q_{DVR}} a_{ki,22}\,||x(n-i)-\beta_k|.x(n) \\ &+\sum_{k=1}^{K_{DVR}}\sum_{i=0}^{Q_{DVR}} a_{ki,23}\,||x(n-i)|-\beta_k|.x(n-i) + \sum_{k=1}^{K_{DVR}}\sum_{i=0}^{Q_{DVR}} a_{ki,24}\,||x(n-i)|-\beta_k|.x(n-i) \end{aligned} \tag{13}$$

where $x(n)$ is I/P and $y(n)$ is O/P.

K_{DVR} is elements in the partition and Q_{DVR} is the memory depth. β_k represents the breakpoint.

5.2. Magnitude Selective Affine (MSA) Based Linearization

Extending the previous work reported for DVR and CPWL, the objective of this work is to further reduce the overheads and complexity of the CPWL method by proposing a magnitude selective affine (MSA) function-based model. The advantage of this method is that it requires only a single linear operation for the selected zone leading to a lower complexity and simpler structure.

In order to optimize the operations, the coefficients in the zone that have similar magnitude can be coupled together [72,73]. The comparison between threshold function and magnitude of input samples can select which zone the samples will be falling and which affine functions can be utilized. This simplification will result in the reduction of the CPWL complex operation. Therefore, we can rewrite the first term of CPWL function in Equation (13) in a simplified way in terms of MSA function that can be expressed as:

$$\begin{aligned} &\textstyle\sum_{m=0}^{M}\sum_{k=0}^{K}\sum_{l=1}^{L} c_{m,k,l}^{(1)}\Big||x(n-k)|^2-\beta_l\Big|x(n-m-k) \\ &=\textstyle\sum_{m=0}^{M}\sum_{k=0}^{K} x(n-m-k)\Big(\sum_{l=1}^{L} c_{m,k,l}^{(1)}\Big||x(n-k)|^2-\beta_l\Big|\Big) \\ &=\textstyle\sum_{m=0}^{M}\sum_{k=0}^{K} u_{m,k}^{(1)}(n-k)x(n-m-k) \end{aligned} \tag{14}$$

$$u_{m,k}^{(1)}(n-k) = \sum_{l=1}^{L} c_{m,k,l}^{(1)}\Big||x(n-k)|^2 - \beta_l\Big| \\ = \begin{cases} A_{m,k,1}^{(1)}|x(n-k)|^2 + B_{m,k,1}^{(1)}, & 0 \le |x(n-k)|^2 < \beta_1 \\ \vdots \\ A_{m,k,L}^{(1)}|x(n-k)|^2 + B_{m,k,L}^{(1)}, & \beta_{L-1} \le |x(n-k)|^2 < \beta_L \end{cases} \tag{15}$$

Here in Equation (15), $A_{m,k,l}^{(1)}$ and $B_{m,k,l}^{(1)}$ are the linear model coefficients defined for each zone of the MSA function $u_{m,k}^{(1)}(.)$. The example of hardware implementation is shown in Figure 9. The simplification shown in Equation (15) leads to this realization that input power terms without any magnitude are compared with the thresholds for the offset and linear gain selection for the MSA function. This leads to removal of square root calculation operation. The overall model of Equation (12) in terms of MSA function can be written as:

$$\begin{aligned} y(n) = & \sum_{m=0}^{M}\sum_{k=0}^{K} u_{m,k}^{(1)}(n-k)x(n-m-k) \\ & + \sum_{m=1}^{M}\sum_{k=0}^{K} u_{m,k}^{(2)}(n-k)x^2(n-k)x^*(n-m-k) \\ & + \sum_{m=1}^{M}\sum_{k=0}^{K} u_{m,k}^{(3)}(n-k)x(n-k)|x(n-m-k)|^2 \\ & + \sum_{m=1}^{M}\sum_{k=0}^{K} u_{m,k}^{(4)}(n-k)x^*(n-k)x^2(n-m-k) \end{aligned} \tag{16}$$

$$u_{m,k}^{(i)}(n-k) = \sum_{l=1}^{L} c_{m,k,l}^{(i)}\Big||x(n-k)|^2 - \beta_l\Big| \\ = \begin{cases} A_{m,k,1}^{(i)}|x(n-k)|^2 + B_{m,k,1}^{(i)}, & 0 \le |x(n-k)|^2 < \beta_1 \\ \vdots \\ A_{m,k,L}^{(i)}|x(n-k)|^2 + B_{m,k,L}^{(i)}, & \beta_{L-1} \le |x(n-k)|^2 < \beta_L \end{cases} \tag{17}$$

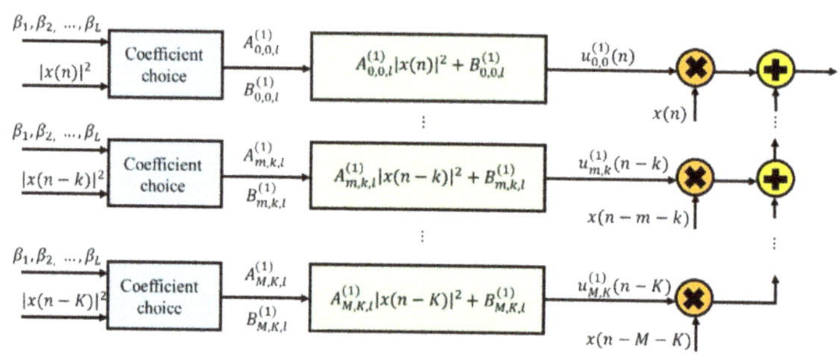

Figure 9. Hardware implementation example using MSA.

5.3. Negative Feedback Iteration based Modelling Approach

For the computation of DPD model coefficients, as shown in the Figure 10 with negative feedback iteration technique, the predistortion coefficients are calculated in the training phase. $z(n)$ is the input of the predistorter, coming through the output of the RoF $y(n)$ here, $z(n) = \frac{y(n)}{G}$ and G is the link's gain. There are two main steps in this technique. The first step is to establish a negative feedback iteration to obtain an input signal that can be regarded as a DPD signal followed by the second step which consists of calculating the DPD model parameters. The shared feedback path is adopted to observe the output information in both frequency bands. In this case, the negative feedback iteration is performed on the lower band and upper band by turns.

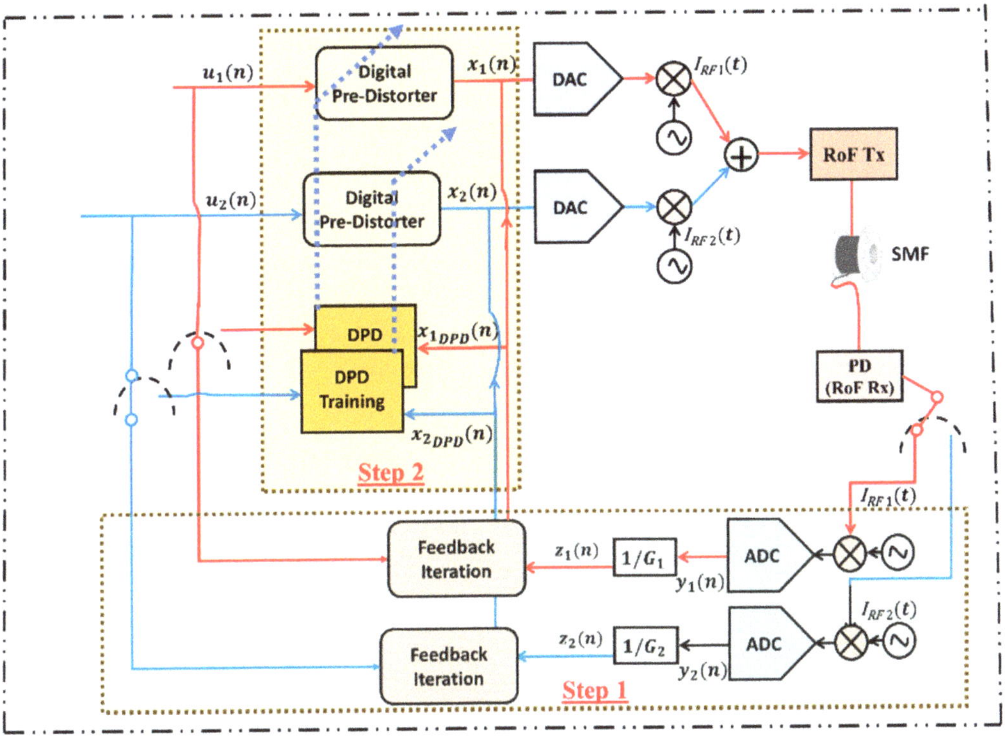

Figure 10. Block diagram representing ILA architecture for Multi Band DPD utilizing Generalized Memory Polynomial (GMP), Canonical Piecewise Linearization (CPWL) and Magnitude Selective Affine (MSA) methodology.

5.4. Estimation Algorithm

Estimation is initiated by the collection of coefficients c_{kq}, d_{kqr} and e_{kqr}, into a R × 1 vector v. here, the total number of coefficients is represented by R and v is related with a signal whose time is sampled over the same period. Coefficients c_{21} are associated to the signal $x(n-1)|x(n-1)|^2$. Z a N × R matrix represents the collection of all such vectors. The predistorter training block output upon convergence will be $z_p(n) = x(n)$ hence, $z(n) = u(n)$.

For N number of samples, output is:

$$z_p = Zv \tag{18}$$

where $Z = [Z(1), \ldots, Z(N)]^T$ and $z_p = [z_p(1), \ldots, z_p(N)]^T$ and v is as mentioned earlier, a R × 1 vector containing set of coefficients c_{kq}, d_{kqr} and e_{kqr}.

The LS solution is the solution for the equation represented by:

$$[Z^H Z]\hat{v} = Z^H x \tag{19}$$

Solution minimizing the cost function is represented by:

$$C = \sum_{n=1}^{N} |z_p(n) - x(n)|^2 \tag{20}$$

6. Experimental Setup

For the validation of this technique, a multiband 5G NR scenario for out/in-door environments working at 3 GHz (20 MHz bandwidth) and 20 GHz (50 MHz bandwidth), which was discussed in our previous work [35], but no DPD was implemented. As an upgradation of this architecture, the setup is integrated with a multiband DPD block to this setup for enhancing the performance of this link. The setup shown in Figure 11 comprises of a 1310 nm optical carrier is modulated by a MZM working with two distinct RF-driven signals and a 1310 nm DFB laser. VSG1 provides RF1 which is a 5G NR waveform at 20 GHz while 5G transceiver provides RF2 which is a 3 GHz flexible (O/G/F-FDM) signal. The process of DPD can be divided in to three main phases.

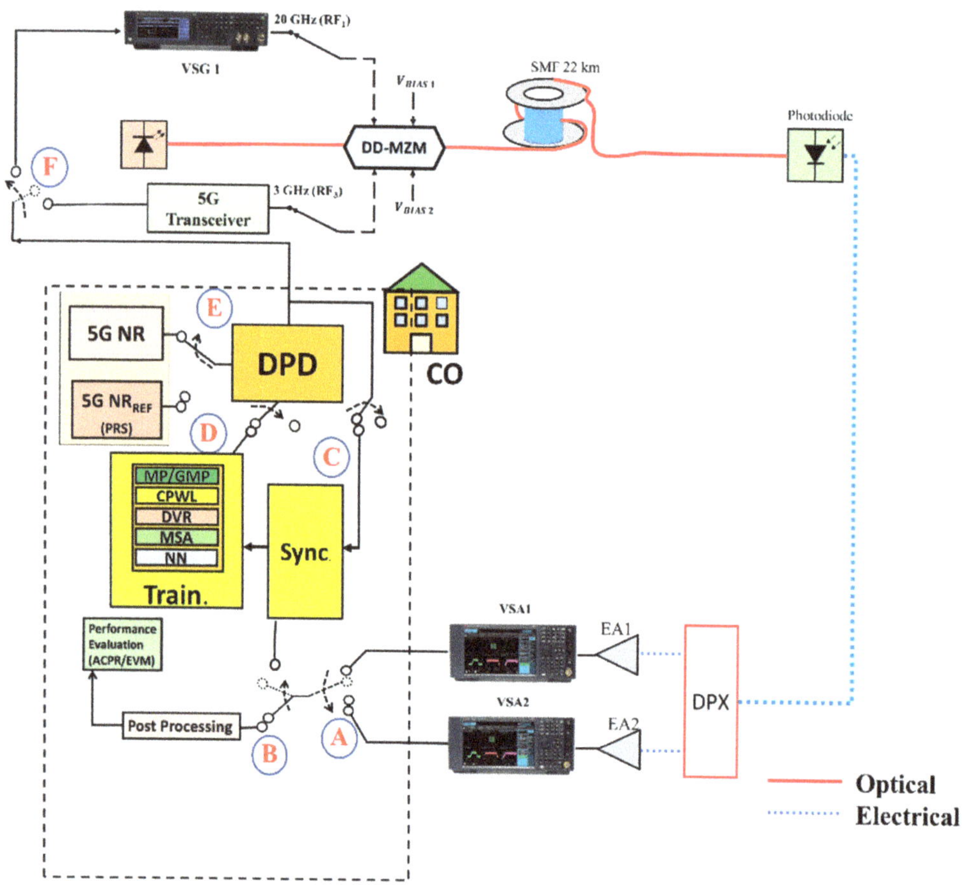

Figure 11. Block Diagram of Experimental Analog Multi Band 5G-NR-OFH RoF. The following Switches have following functionalities: A: Selection between VSA 1 and 2. B: Selection between post processing block for performance or Synchronization block. C: Connected to Synch. Block for performing synchronization followed by DPD training. D: DPD blocks connection with training DPD block E: Required for time synchronization (TS) procedure. F: Required for performing validation to launch DPD inputs to the VSGs. For DPD training phase, architectures can be selected that are requires such as MSA, CPWL, ONN, etc.

In the first phase, we upconvert these signals at the respective carrier frequencies of 3 and 20 GHz, respectively, one after the other followed by the passing of these signals through 22 km of Standard Single-Mode Fiber (SSMF) and photodetector (0.71 A/W and

40 GHz bandwidth) receives the signal and converts the received signal back to electrical domain. Since, the multiband needs to be isolated separately, an amplification stage is added. Followed by a diplexer (DPX) that separates the 20 GHz and 3 GHz signals. The signals then go to distinct vector signal analyzers (VSA). Here each VSA outputs is fed to the post processing block for performance evaluation. This step is carried out without DPD, this means that the output is evaluated without DPD process in this step.

In the second phase called as DPD training phase, the DPD operation depicted in the Figure 10 is utilized in this section and training is employed unless the error converges.

In simple words, DPD ensures that the phase and amplitude responses are inverse to that obtained at electrical amplifiers EA1 and EA2, respectively. The architectures such as GMP, CPWL, MSA and ONN can be utilized as per the user requirement and comparative requirements.

In order to achieve synchronization for the received input and the output waveform, we exploit PRS (Position reference signal) presented in the 5G NR framework. The bandwidth for PRS is taken to be 20 MHz/106 resource blocks. The received and output reference transmitted signal are correlated in time domain and the PDP (power delay profile) is passed through the maximum block to find the strongest path of arrival.

Moving on to the third phase, the predistorted basebands signal are passed into the DPD block upconverting the signals at their carrier frequency by their respective VSGs after which they are fed into the optical link. Then, the signal that is received at photodiode is passed through diplexer DPX to isolate the respective multi-bands ate the VSA passes to the DPD training stage. In the phase where DPD validation is carried out, we move the switches to the opposite direction. The evaluation for 5G NR frames is achieved which involves predistortion followed by passing them to the VSG. The RoF link's nonlinearities slowly fluctuate owing to thermal effects and component ageing, from which we deduce that real-time processing in the adaptation is inessential. The parameters utilized are summarized in Table 5 that have been used previously in [35] and other state of the art [20–28].

Table 5. Optical Link Parameters.

Parameters	Values
5G NR Waveforms	f_c = 3 and 20 GHz Flexible G/F-OFDM Constellation type = 256 QAM
Laser	Wavelength = 1310 nm Transmitter Type = Mach Zehnder Modulator
Optical Fiber	Type = SSMF Fiber Dispersion = 16 $\frac{\text{ps}}{\text{nmkm}}$ Fiber Distance = 22 km Attenuation = 0.42 $\frac{\text{dB}}{\text{km}}$
Photodiode	Responsivity = 0.71 A/W Bandwidth = 40 GHz

The parameters of different architectures that are utilized in this study for the experimental setup are given in Tables 6–8, respectively. Table 6 enlists the parameters of the NN that results in the optimized performance of NN. The parameters discussed previously in Section 3.1 are utilized to define the structure of the proposed Optimized Neural Network and their values are mentioned in the Table 6. The last section of the table evaluates the complexity aspect of the NN by evaluating its expressions in terms of its coefficients. For a comparative analysis, the parameters of GMP and DVR are also mentioned below in Table 7. The proposed MSA-DPD technique and CPWL without modification is used with $M = 3$ and $K = L = 4$. Similarly, for comparison, we have used GMP method previously used in [34,65] with the parameters $K = Q = R = 3$. For NN, $N = 10$, $K = 30$ is utilized.

Table 6. ONN parameters.

Architecture	Parameters
Component	**Value/Specification**
Optimizer	ADAM
Activation function	ReLu
O/P layer activation function	Softmax
Loss- function	Mean Square Error (MSE)
No. of Hidden-Layers N	10
No. of Neurons per layer K	30
Component	**Value/Specification**
Regularization method	Dropout
Regularization factor	0.001
Learning rate	0.01
Batch size	16, 32, 64, 128, 256, 512, 1024
Number of epochs	100
Validation split	0.4
Training	
Component	**Value/Specification**
Number of training samples	1,000,000
Number of testing samples	1,000,000
Complexity	
Equation	**Value**
$(N-1)\,K^2 + (4+N)\,K + 6$	8526

Table 7. GMP and DVR parameters.

GMP	Parameters
Component	**Value**
Nonlinearity Order (K)	3
Memory depth (Q)	3
Delay taps length (R)	3
DVR	
Component	**Value**
Elements in partition (K)	3
Memory depth (Q)	3
Breakpoint (β)	1

Table 8. CPWL and MSA parameters.

CPWL	Value
Memory Depth (M)	3
Number of partitions (L)	4
Length of the FIR filter (K)	4
MSA	**Value**
Memory Depth (M)	3
Number of partitions (L)	4
Length of the FIR filter (K)	4

7. Experimental Results and Discussion

In this section, the results are discussed for the experimental setup discussed in the earlier section. The Mean Square Error (MSE) is one of the ways to estimate the accuracy of the estimation of coefficients for different architectures utilized. The MSE when No DPD is applied is measured to be −27 dB, while for GMP it is −30 dB. The value reaches as low as −35 dB for CPWL and MSA while NN has a MSE of −39 dB.

In addition to MSE, the wellness of the proposed methods is compared and are presented in form of Adjacent Channel Leakage or Power Ratio (ACLR/ACPR) and Error Vector Magnitude (EVM).

7.1. Error Vector Magnitude

Error Vector Magnitude is the most common performance metrics that is utilized for performance evaluation for this study item in 3GPP. EVM determines the difference between the 'expected' value of the symbol in demodulated form to 'actual' value of the demodulated received symbol. EVM can be mathematically written as [5]:

$$EVM\ (\%) = \sqrt{\frac{\frac{1}{M}\sum_{m=1}^{M} |S_m - S_{0,m}\ |^2}{\frac{1}{M}\sum_{m=1}^{M} |S_m\ |^2}} \tag{21}$$

where M is the quantity of constellation symbols, S_m is the real symbol of the constellation associated with the symbol "m" and $S_{0,m}$ is the real symbol associated with S_m. The 3GPP has set an EVM limit for 256 QAM to be 3.5% [74].

In Figure 12a, the Error Vector Magnitude EVM is represented by sweeping the RF input power. It is evident that MSA-DPD results in EVM reduction to <3% as compared to 5% obtained with GMP. The MSA-DPD has a slight improvement as compared to CPWL, but this is not the significant contribution, we expect to have similar improvement but with smaller complexity. In addition to this, DPD NN has a better performance as compared to MSA-DPD by 1%. It can be seen that DPD NN has a better performance as compared to MSA technique between 0 and 5 dB, however, MSA has overall better performance from −15 to 0 dB as compared to all other techniques. Similarly, In Figure 12b, EVM is reported for all the comparative methodologies employed for different flexible waveforms that has been used in 5G NR framework for 0 dBm RF input power. Clearly, MSA-DPD has almost similar performance as compared to CPWL knowing MSA reduces the complexity as compared to CPWL. It can be seen that MSA has a better reduction by 1% as compared to CPWL. When compared to NN, the NN has better performance as compared to MSA. It is clearly observable that NN has the best performance.

7.2. Adjacent Channel Leakage Ratio

ACLR also called as Adjacent Channel Power Ratio (ACPR) is a quantity that determines the distortion components outside the useful signal bandwidth. It is expressed as [5]:

$$ACLR_{dBc} = 10\log_{10}\left[\frac{\int_{ab_l}^{ab_u} T(f)\ df}{\int_{ub_l}^{ub_u} T(f) df}\right] \tag{22}$$

where $T(f)$ denotes Power Spectral Density (PSD) of the output signal while ab_u and ab_l are the upper and lower frequency limits of the adjacent channel; ub_l and ub_u are the frequency bounds of useful bands.

In addition to EVM, we look at the ACLR also called as MIA. In Figure 13a, ACLR for given input power variations and for different methodologies is represented. We can see that, at an RF I/P power of 0 dBm, the ACPR value for no DPD utilization is around −28 dBc, for DPD-GMP the value is around −41 dBc, for DPD-CPWL, it is around −44 dBc, for MSA-DPD, it is around −45 dBc and finally for DPD-NN, the value is around −48 dBc keeping the ACPR below −45 dBc set by 3GPP [74]. With this it is obvious that DPD-NN has achieved significantly 3 dBc lower ACPR value as compared to MSA-DPD which has the best performance as compared to other Volterra series methods.

Similarly, in Figure 13b, the spectral density is shown. The power spectral density (PSD) with and without DPD at 3 GHz is shown. It is observed that MSA technique is achieving good performance as compared to CPWL while the NN DPD has a little advantage over the MSA with spectral regrowth suppression in the 10 to 15 MHz zones.

Similarly, in Figure 13c, the electrical spectra are shown, there are two main components that show the carrier signals at 3 and 20 GHz, respectively. It represents that the aim of the proposed MSA-DPD technique further reduces the complexity and implementation of the DPD methodology and enhance the performance of the link as compared to CPWL, GMP and when no DPD is applied. Therefore, it is observable that ACPR is reduced with the proposed MSA-DPD methods as compared to CPWL and GMP method by a good proportion keeping the ACPR below −45 dBc set by 3GPP [74]. It is important to observe that MSA-DPD performs a little better than CPWL, however, the performance gain is not the only benefit, but the complexity reduction is the most important benefit of the proposed MSA-DPD technique (discussed in Section 7.3).

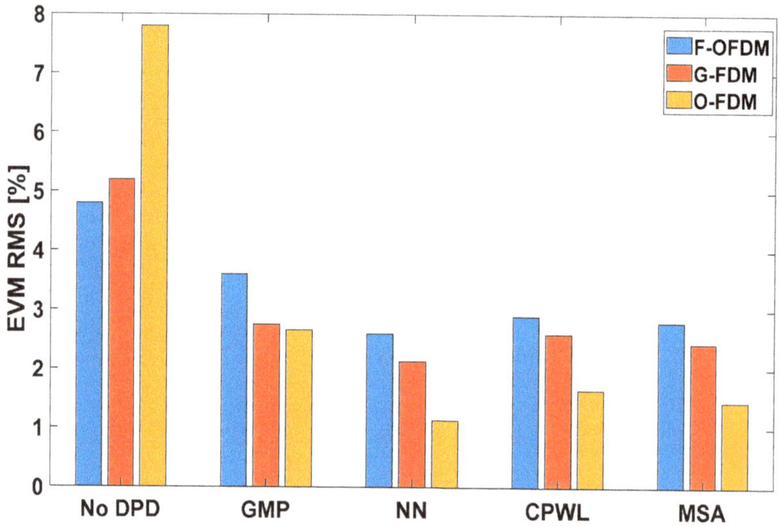

(a)

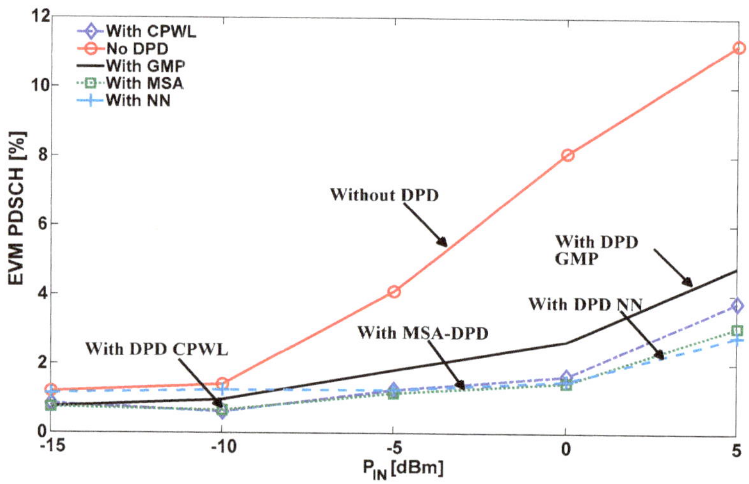

(b)

Figure 12. EVM results (**a**) for DPD methods utilized in this study vs. No DPD case. DPD with CPWL (blue), GMP (black), MSA (green), NN (light blue) and No DPD (red). (**b**) 5G NR OFH performance DPD efficacy with MSA, CPWL, GMP and without DPD for flexible 5G transceiver waveforms.

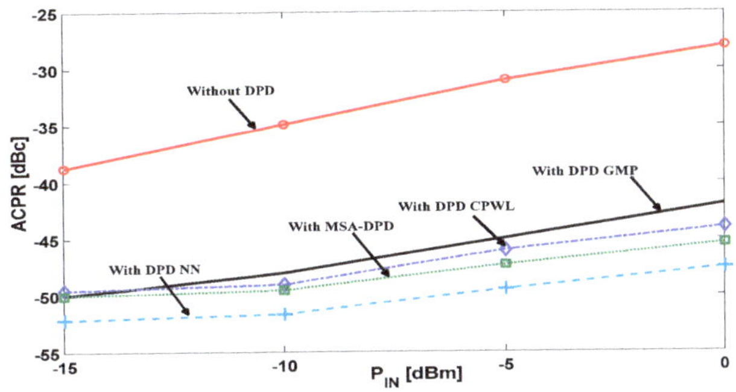

(**a**)

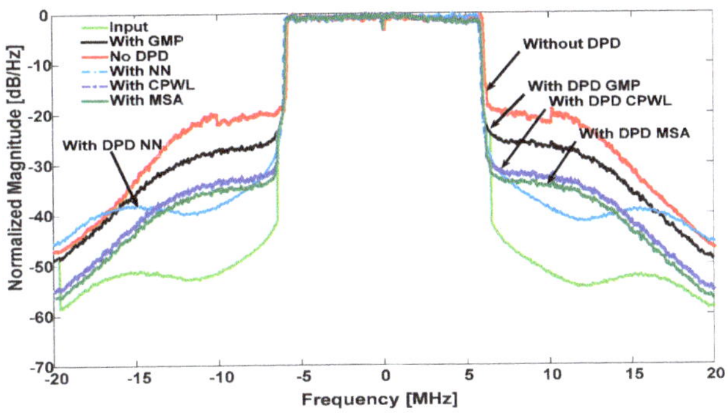

(**b**)

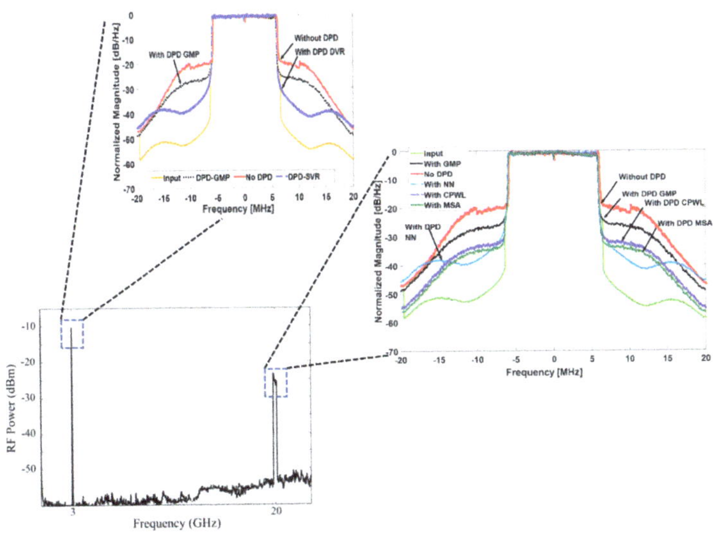

(**c**)

Figure 13. 5G NR OFH performance (**a**) ACPR vs. varying RF input power. (**b**) DPD efficacy with MSA, CPWL, GMP and without DPD in terms of Power Spectral Density/Spectral Regrowth. (**c**) Electrical Spectra showing the two carrier frequencies.

7.3. Complexity Considerations

The complexity reduction that MSA-DPD brings with gaining similar performance as compared to CPWL method is significant contribution. The complexity of the DPD model construction of the models is shown in Table 9, which is mainly measured by the required number of real multipliers, as the multipliers take up most of the hardware resources. Table 9 signifies that MSA DPD (220 multiplications) has much lesser complexity as compared to CPWL (880 multiplications). The advanced variations in Volterra series can be obtained by changing memory depth and nonlinearity order to higher numbers. However, the computational complexity has to be considered as shown in Equation (11). This means that while selecting the DPD model and its complexity, a smart trade-off between complexity and performance can be made accordingly. For a comparative evaluation of NN and GMP methods in terms of complexity and performance, we evaluate expressions for each method in term of its complexity. The number of coefficients and multiplications in CPWL architecture are higher as compared to GMP, however, the complexity in CPWL results in better performance as compared to GMP (as seen in results section in Figures 11 and 12). The reduction in complexity is brought into the play as proposed in MSA method. The performance of MSA achieves similar performance as compared to CPWL, however, the complexity in terms of coefficients in MSA is reduced from 520 to 260 and multiplications are reduced from 880 to 220. Similarly, looking at NN architecture, the performance (see Figures 11 and 12) is comparable to MSA which is in similar proportions to NN, however, the number of coefficients rise to 8526 and number of multiplications reached to 8224. NN complexity is a challenging issue, and this can be reduced as shown in this work that a very limited number of hidden layers and neuron per layer are employed. The multiplications and complexity of the NN increases exponentially with the increase in number of hidden layers and neurons per layer, depending on the application, MSA with much less complexity and similar based DPD can be employed as they achieve a similar performance as compared to NN.

Table 9. Complexity Comparisons.

DPD Method	Coefficients	# Coefficients	Multiplications
GMP	$K_a(Q_a+1)+K_b(Q_b+1)R_b+K_c(Q_c+1)R_c$	84	244
CPWL	$(4M+1)(K+1)L$	260	$(14M+2)(K+1)L = 880$
MSA	$2(4M+1)(K+1)L$	520	$(14M+2)(K+1) = 220$
NN	$(N-1)K^2+(4+N)K+6$	8526	$(N-1)K^2+4K+4 = 8224$

7.4. Real Time Implementation

In a realistic NN scenario, linearization methodology is carried out at the Central Office (CO) where the BBUs are placed and a periodical re-training of the DPD system is in this case necessary, requiring however a negligible time with respect to the time of normal operation of the RoF system. Recently, a Xilinx DPD kit has been developed that can be used for this purpose [75].

Bringing feedback signal from the BS to RAU is one of the main challenging task in the adaptive recompense of the RoF link. This is due to possible nonlinearity of the feedback link; actually, it can be as nonlinear as the RoF link, which is being compensated for. The present work is based on the fact that the predistorter sees only the non-idealities that it needs to compensate. Since it is assumed that the nonlinear feedback connection is uncompensated, therefore, it would destroy the performance of the compensation. In simple words, it means that an approach is utilized where the RoF link is first compensated for using a post distorter and known training signal from the RAU is used here. After that, the already compensated for downlink RoF link can be used as a feedback connection for the compensation [76]. However, as a possible feedback scenario, Figure 14 represents a realistic implementation of an adaptive DPD that shifts complex processing of the signals at the Central Office/Base Transmit Station (CO-BTS).

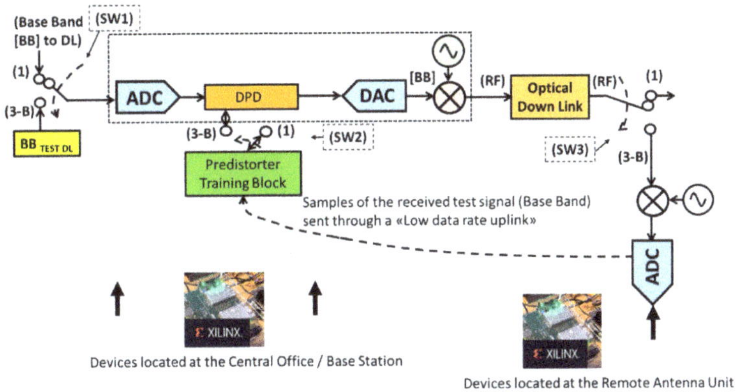

Figure 14. Conceptual Possible Realization of an Adaptive Predistortion Scheme.

It should be noted that DPD works as a black box, it counter acts the overall nonlinearities of the system including that of MZM (laser), fiber and photodiode. For tens of km, the combined effect of laser chirp and fiber dispersion becomes a major nonlinearity issue after tens of km [5]. Therefore, laser and possibly photodiode are the primary source of nonlinearity which is mitigated in this proposed bench. In future, it will be interesting to increase the length of fiber and linearize the link by mitigating the fiber nonlinearities such as Kerr effect.

Indeed, with the higher modulation format and higher bandwidth similar to multiple LTE carriers or 5G new radio (NR) waveforms as discussed, they would lead to higher complexity of DPD operation due to stronger PAPR. Concomitantly, the elevation in bandwidth will lead to overall increase in the base-band memory of the system model. Nevertheless, the evaluated models are still valid. However, higher values of the Q and K will be indispensable as compared to the considered case.

To summarize the discussion above, Table 10 lists the values of ACPR, MSE and EVM for all the utilized methodologies.

Table 10. Results Summary for ACPR (0 dBm), EVM (5 dBm) and MSE.

Methodology	ACPR (dBc)	EVM (%)	MSE (dB)
No-DPD	−28	11	−27
DPD-GMP	−41	5	−30
DPD-CPWL	−44	4	−35
DPD-MSA	−45	3.5	−35
DPD-ONN	−48	2.7	−39

8. Conclusions

This work summarizes a successful realization of 5G NR multiband OFH with a novel unprecedented DPD solutions for reducing RoF nonlinearities using the conventional volterra based methods and further exploiting deep learning methods to further improve results using Neural Networks. Firstly, a novel MSA DPD method has been proposed which reduces the complexity of CPWL method and reaches similar performance with reduction of complexity overheads and multiplications by 75%. The article also explains the theoretical foundations and elements required for building a Neural Network. The 5G NR multiband signals at 3 GHz and 20 GHz are employed to 22 km fiber length. The proposed MSA-DPD method results in reduction of ACPR from −28 dBc to −45 dBc while NN results in ACPR reduction to −48 dBc. Similarly, 11% of EVM is reduced to 3.5% with MSA method and NN leads to 2.7% at RF input power of 5 dBm. The results signify that proposed MSA-DPD method reduces the signal impairments in better proportions as compared to GMP method and CPWL. The estimated multiplication operations from CPWL to MSA are reduced from 880 to 220 leading to much less complexity and overheads meeting the standardization requirements set by 3GPP Release 17. MSA with much less complexity and similar performance to NN can be employed for DPD depending on the application scenario.

Author Contributions: Conceptualization, M.U.H.; methodology, M.U.H. and J.S.; software, M.U.H. and M.R.; validation, M.I.A., M.A. and M.U.H.; formal analysis, M.U.H.; investigation, M.U.H. and M.A.; resources, M.I.A. and M.U.H.; data curation, J.S.; writing—original draft preparation, M.U.H. and M.R.; writing—review and editing, M.U.H., M.A., M.R., M.I.A., J.S.; visualization, J.S.; supervision, M.U.H. and M.I.A.; project administration, M.A.; funding acquisition, M.R. and M.U.H. All authors have read and agreed to the published version of the manuscript.

Funding: This research received no external funding.

Institutional Review Board Statement: Not applicable.

Data Availability Statement: The data presented in this study are available on request from the corresponding author.

Acknowledgments: Authors would like to thank Isha Mittal and Samridhi Mishra for their efforts in experimentation and data analysis.

Conflicts of Interest: The authors declare no conflict of interest.

References

1. Hadi, M.; Awais, M.; Raza, M.; Khurshid, K.; Jung, H. Neural Network DPD for Aggrandizing SM-VCSEL-SSMF-Based Radio over Fiber Link Performance. *Photonics* **2021**, *8*, 19. [CrossRef]
2. 3GPP, User Equipment (UE) Radio Transmission and Reception; Part 3: Range 1 and Range 2 Interworking Operation with Other Radios: 2019, TS 38.101-3 Version 16.0.0 Release 16. Available online: https://www.etsi.org/deliver/etsi_ts/138100_138199/13810103/16.06.00_60/ts_13810103v160600p.pdf (accessed on 5 November 2021).
3. Nanni, J.; Polleux, J.-L.; Algani, C.; Rusticelli, S.; Perini, F.; Tartarini, G. VCSEL-based radio-over-G652 fiber system for short−/medium-range MFH solutions. *IEEE/OSA J. Lightwave Technol.* **2018**, *36*, 4430–4437. [CrossRef]
4. Khurshid, K.; Khan, A.A.; Siddiqui, H.; Rashid, I.; Hadi, M.U. Big Data Assisted CRAN Enabled 5G SON Architecture. *J. ICT Res. Appl.* **2019**, *13*, 93–106. [CrossRef]
5. Hadi, M.U.; Nanni, J.; Polleux, J.-L.; Traverso, P.A.; Tartarini, G. Direct digital predistortion technique for the compensation of laser chirp and fiber dispersion in long haul radio over fiber links. *Opt. Quantum Electron.* **2019**, *51*, 205. [CrossRef]
6. Hadi, M.U.; Murtaza, G. Enhancing distributed feedback-standard single mode fiber-radio over fiber links performance by neural network digital predistortion. *Microw. Opt. Technol. Lett.* **2021**, *63*, 1558–1565. [CrossRef]
7. Zhang, X.; Zhu, R.; Shen, D.; Liu, T. Linearization Technologies for Broadband Radio-Over-Fiber Transmission Systems. *Photonics* **2014**, *1*, 455–472. [CrossRef]
8. Li, H.; Bauwelinck, J.; Demeester, P.; Torfs, G.; Verplaetse, M.; Verbist, J.; Van Kerrebrouck, J.; Breyne, L.; Wu, C.-Y.; Bogaert, L.; et al. Real-Time 100-GS/s Sigma-Delta Modulator for All-Digital Radio-Over-Fiber Transmission. *J. Lightwave Technol.* **2020**, *38*, 386–393. [CrossRef]
9. Hadi, M.U.; Jung, H.; Ghaffar, S.; Traverso, P.A.; Tartarini, G. Optimized digital radio over fiber system for medium range communication. *Opt. Commun.* **2019**, *443*, 177–185. [CrossRef]
10. Wang, J.; Jia, Z.; Campos, L.A.; Knittle, C.; Jia, S. Delta-Sigma Modulation for Next Generation Fronthaul Interface. *J. Lightwave Technol.* **2018**, *37*, 2838–2850. [CrossRef]
11. Hadi, M.U.; Traverso, P.A.; Tartarini, G.; Jung, H. Experimental characterization of Sigma Delta Radio over fiber system for 5G C-RAN downlink. *ICT Express* **2020**, *6*, 23–27. [CrossRef]
12. Hadi, M.U.; Jung, H.; Traverso, P.A.; Tartarini, G. Experimental evaluation of real-time sigma-delta radio over fiber system for fronthaul applications. *Int. J. Microw. Wirel. Technol.* **2020**, *13*, 756–765. [CrossRef]
13. Hadi, M.U.; Kausar, S.; Mittal, I. Efficient Pre-BPF based Sigma Delta Radio over Fiber System for 5G NR Fronthauls. In Proceedings of the IEEE EUROCON 2021—19th International Conference on Smart Technologies, Lviv, Ukraine, 6–8 July 2021; pp. 308–311.
14. Draa, M.N.; Hastings, A.S.; Williams, K.J. Comparison of photodiode nonlinearity measurement systems. *Opt. Express* **2011**, *19*, 12635–12645. [CrossRef] [PubMed]
15. Chen, Z.; Yan, L.; Pan, W.; Luo, B.; Zou, X.; Guo, Y.; Jiang, H.; Zhou, T. SFDR enhancement in analog photonic links by sim-ultaneous compensation for dispersion and nonlinearity. *OSA Opt. Expr.* **2013**, *21*, 20999–21009. [CrossRef]
16. Haas, B.M.; Murphy, T. A Simple, Linearized, Phase-Modulated Analog Optical Transmission System. *IEEE Photonics Technol. Lett.* **2007**, *19*, 729–731. [CrossRef]

17. Hraimel, B.; Zhang, X. Characterization and compensation of AM-AM and AM-PM distortion in mixed polarization radio over fiber systems. In Proceedings of the 2012 IEEE MTT-S International Microwave Symposium (IMS), Montreal, QC, Canada, 1–3 June 2012.
18. Zhu, R.; Zhang, X. Linearization of Radio-Over-Fiber Systems by Using Two Lasers with Different Wavelengths. In Proceedings of the 2014 IEEE MTT-S International Microwave Symposium (IMS), Tampa, FL, USA, 1–6 June 2014.
19. Ghannouchi, F.; Younes, M.; Rawat, M. Distortion and impairments mitigation and compensation of single- and multi-band wireless transmitters. IET Microw. *Antennas Propag.* **2013**, *7*, 518–534.
20. Duan, R.; Xu, K.; Dai, J.; Cui, Y.; Wu, J.; Li, Y.; Dai, Y.; Li, J. Linearity improvement based on digital signal processing in in-tensity-modulated analog optical links incorporating photonic frequency down conversion. In Proceedings of the Optical Fiber Communication Conference and Exposition, Los Angeles, LA, USA, 6–8 March 2012.
21. Vieira, L.; Gomes, N.J.; Nkansah, A.; van Dijk, F. Behavioral modeling of radio-overfiber links using memory polynomials. In Proceedings of the 2010 IEEE Topical Meeting on Microwave Photonics (MWP), Montreal, QC, Canada, 5–9 October 2010; pp. 85–88.
22. Pei, Y.; Xu, K.; Li, J.; Zhang, A.; Dai, Y.; Ji, Y.; Lin, J. Complexity-reduced digital predistortion for subcarrier multiplexed radio over fiber systems transmitting sparse multi-band RF signals. *Opt. Express* **2013**, *21*, 3708–3714. [CrossRef]
23. Hekkala, A.; Hiivala, M.; Lasanen, M.; Perttu, J.; Vieira, L.C.; Gomes, N.J.; Nkansah, A. Predistortion of Radio Over Fiber Links: Algorithms, Implementation, and Measurements. *IEEE Trans. Circuits Syst. I Regul. Pap.* **2011**, *59*, 664–672. [CrossRef]
24. Fuochi, F.; Hadi, M.U.; Nanni, J.; Traverso, P.A.; Tartarini, G. Digital predistortion technique for the compensation of nonlinear effects in radio over fiber links. In Proceedings of the 2016 IEEE 2nd International Forum on Research and Technologies for Society and Industry Leveraging a Better Tomorrow (RTSI), Bologna, Italy, 7–9 September 2016; pp. 1–6.
25. Vieira, L.C.; Gomes, N.J.; Nkansah, A. An experimental study on digital predistortion for radio-over-fiber links. *Asia Commun. Photonics Conf. Exhib.* **2010**, 126–127. [CrossRef]
26. Hekkala, A.; Lasanen, M.; Vieira, L.C.; Gomes, N.J.; Nkansah, A. Architectures for Joint Compensation of RoF and PA with Nonideal Feedback. In Proceedings of the 2010 IEEE 71st Vehicular Technology Conference, Taipei, Taiwan, 16–19 May 2010; pp. 1–5.
27. Mateo, C.; Carro, P.L.; Garcia-Ducar, P.; De Mingo, J.; Salinas, I. Experimental evaluation of the feedback loop effects in digital predistortion of a radio-over-fiber system. In Proceedings of the 2017 Ninth International Conference on Ubiquitous and Future Networks (ICUFN), Milan, Italy, 4–7 July 2017; pp. 1039–1041.
28. Mateo, C.; Clemente, J.; Garcia-Ducar, P.; Carro, P.L.; de Mingo, J.; Salinas, I. Digital predistortion of a full-duplex Radio-over-Fiber mobile fronthaul link with feedback loop. In Proceedings of the 2017 13th International Wireless Communications and Mobile Computing Conference (IWCMC), Valencia, Spain, 26–30 June 2017; pp. 1425–1430.
29. Mateo, C.; Carro, P.L.; García-Dúcar, P.; De Mingo, J.; Salinas, Í. Minimization of Feedback Loop Distortions in Digital Predistortion of a Radio-Over-Fiber System with Optimization Algorithms. *IEEE Photonics J.* **2017**, *9*, 7904414. [CrossRef]
30. Roselli, L.; Borgioni, V.; Zepparelli, F.; Ambrosi, F.; Comez, M.; Faccin, P.; Casini, A. Analog laser predistortion for multiservice radio-over-fiber systems. *J. Lightwave Technol.* **2003**, *21*, 1211–1223. [CrossRef]
31. Aslam, N.; Xia, K.; Hadi, M.U. Optimal Wireless Charging Inclusive of Intellectual Routing Based on SARSA Learning in Renewable Wireless Sensor Networks. *IEEE Sens. J.* **2019**, *19*, 8340–8351. [CrossRef]
32. Hraimel, B.; Zhang, X. Low-cost broadband predistortion-linearized singledrive x-cut Mach-Zehnder modulator for ra-dio-over-fiber systems. *Photonics Technol. Lett.* **2012**, *24*, 1571–1573. [CrossRef]
33. Hadi, M.U.; Khurshid, K.; Basit, A. Pre-BPF based Sigma Delta Radio over Fiber for 5G Sub-6 GHz Fronthauls. In Proceedings of the 2021 International Conference on Communication Technologies (ComTech), Washington, DC, USA, 20–25 June 2021; pp. 13–16. [CrossRef]
34. Hadi, M.U.; Nanni, J.; Traverso, P.A.; Tartarini, G.; Venard, O.; Baudoin, G.; Polleux, J.-L. Experimental evaluation of digital predistortion for VCSEL-SSMF-based Radio-over-Fiber link. In Proceedings of the 2018 International Topical Meeting on Microwave Photonics (MWP), Toulouse, France, 22–25 October 2018; pp. 1–4.
35. Hadi, M.U.; Kantana, C.; Traverso, P.A.; Tartarini, G.; Venard, O.; Baudoin, G.; Polleux, J. Assessment of digital predistortion methods for DFB-SSMF radio-over-fiber links linearization. *Microw. Opt. Technol. Lett.* **2020**, *62*, 540–546. [CrossRef]
36. Hadi, M.U.; Awais, M.; Raza, M. Multiband 5G NR-over-Fiber System Using Analog Front Haul. In Proceedings of the 2020 International Topical Meeting on Microwave Photonics (MWP), Matsue, Japan, 24–26 November 2020; pp. 1–4.
37. James, J.; Shen, P.; Nkansah, A.; Liang, X.; Gomes, N.J. Nonlinearity and Noise Effects in Multi-Level Signal Millimeter-Wave Over Fiber Transmission Using Single and Dual Wavelength Modulation. *IEEE Trans. Microw. Theory Tech.* **2010**, *58*, 3189–3198. [CrossRef]
38. Jeon, D.-H.; Jung, H.-D.; Han, S.-K. Mitigation of dispersion-induced effects using SOA in analog optical transmission. *IEEE Photonics Technol. Lett.* **2002**, *14*, 1166–1168. [CrossRef]
39. Usman Hadi, M.; Mittal, I.; Siraj, M. 5G NR Fiber Wireless System using Analog Optical Front Haul for Long Reach Applications. In Proceedings of the 2021 International Conference on Communication, Control and Information Sciences (ICCISc), Online, 24–25 September 2021; pp. 1–4. [CrossRef]
40. Ismail, T.; Liu, C.-P.; Mitchell, J.; Seeds, A.J. High-Dynamic-Range Wireless-Over-Fiber Link Using Feedforward Linearization. *J. Lightwave Technol.* **2007**, *25*, 3274–3282. [CrossRef]

41. Park, S.-H.; Choi, Y.-W. Significant suppression of the third intermodulation distortion in transmission system with optical feedforward linearized transmitter. *IEEE Photonics Technol. Lett.* **2005**, *17*, 1280–1282. [CrossRef]
42. Korotky, S.K.; De Ridder, R. Dual parallel modulation schemes for low-distortion analog optical transmission. *IEEE J. Sel. Areas Commun.* **1990**, *8*, 1377–1381. [CrossRef]
43. Gamage, P.A.; Nirmalathas, A.; Lim, C.; Novak, D.; Waterhouse, R. Design and Analysis of Digitized RF-Over-Fiber Links. *J. Lightwave Technol.* **2009**, *27*, 2052–2061. [CrossRef]
44. Zhang, L.; Pang, X.; Ozolins, O.; Udalcovs, A.; Popov, S.; Xiao, S.; Hu, W.; Chen, J. Spectrally efficient digitized radio-over-fiber system with k-means clustering-based multidimensional quantization. *Opt. Lett.* **2018**, *43*, 1546–1549. [CrossRef]
45. Shiina, R.; Fujiwara, T.; Taniguchi, T.; Ikeda, S. DRoF-based optical video re-transmission of commercial RF broadcast signals. *J. Opt. Commun. Netw.* **2019**, *11*, 559–567. [CrossRef]
46. Mathe, D.M.; Neto, B.; Oliveira, R.S.; Teixeira, A.L.J.; Costa, J.C.W.A. Adaptive Signal Compression in Intermediate Frequency Band for DRoF Mobile Fronthaul. *IEEE Photonics Technol. Lett.* **2019**, *31*, 579–582. [CrossRef]
47. Hadi, M.U. Sigma Delta Radio Frequency Transport System for 5G Sub-6 GHz Fronthauls. In Proceedings of the 2020 IEEE 15th International Conference on Industrial and Information Systems (ICIIS), Rupnagar, India, 26–28 November 2020; pp. 596–600.
48. Sezgin, I.C.; Dahlgren, M.; Eriksson, T.; Coldrey, M.; Larsson, C.; Gustavsson, J.; Fager, C. A Low-Complexity Distributed-MIMO Testbed Based on High-Speed Sigma–Delta-Over-Fiber. *IEEE Trans. Microw. Theory Tech.* **2019**, *67*, 2861–2872. [CrossRef]
49. Hadi, M.U.; Basit, A.; Khurshid, K. MIMO Enabled MultiBand 5G NR FiberWireless using Sigma Delta over FiberTechnology. In Proceedings of the 2021 International Bhurban Conference on Applied Sciences and Technologies (IBCAST), Islamabad, Pakistan, 12–16 January 2021; pp. 1007–1010.
50. Cui, Y.; Zhang, M.; Wang, D.; Liu, S.; Li, Z.; Chang, G.K. Bit-based support vector machine nonlinear detector for millime-ter-wave radio-over-fiber mobile fronthaul systems. *Opt. Express* **2017**, *25*, 26186–26197. [CrossRef]
51. Hadi, M.U.; Mittal, I. On the Use of SVR based Machine Learning Method for Nonlinearities Mitigation in Short Range Fronthaul Links. In Proceedings of the 2021 10th IEEE International Conference on Communication Systems and Network Technologies (CSNT), Bhopal, India, 18–19 June 2021; pp. 628–631.
52. Gonzalez, N.G.; Zibar, D.; Caballero, A.; Monroy, I.T. Experimental 2.5-Gb/s QPSK WDM Phase-Modulated Radio-Over-Fiber Link with Digital Demodulation by a *K*-Means Algorithm. *IEEE Photonics Technol. Lett.* **2010**, *22*, 335–337. [CrossRef]
53. Lee, J.; He, J.; Wang, Y.; Fang, C.; Wang, K. Experimental demonstration of millimeter-wave radio-over-fiber system with convolutional neural network and binary convolutional neural network (BCNN). *arXiv* **2020**, arXiv:2001.02018.
54. Lee, J.; He, J.; Wang, K. Neural networks and FPGA hardware accelerators for millimeter-wave radio-over-fiber systems. In Proceedings of the International Conference on Transparent Optical Networks, Bari, Italy, 19–23 July 2020; p. Mo.D1.5.
55. Liu, S.; Wang, X.; Zhang, W.; Shen, G.; Tian, H. An Adaptive Activated ANN Equalizer Applied in Millimeter-Wave RoF Transmission System. *IEEE Photonics Technol. Lett.* **2017**, *29*, 1935–1938. [CrossRef]
56. Liu, S.; Xu, M.; Wang, J.; Lu, F.; Zhang, W.; Tian, H.; Chang, G.-K. A Multilevel Artificial Neural Network Nonlinear Equalizer for Millimeter-Wave Mobile Fronthaul Systems. *J. Lightwave Technol.* **2017**, *35*, 4406–4417. [CrossRef]
57. Liu, S.; Alfadhli, Y.M.; Shen, S.; Tian, H.; Chang, G.K. Mitigation of multi-user access impairments in 5G A-RoF-based mobile-fronthaul utilizing machine learning for an artificial neural network nonlinear equalizer. In Proceedings of the Optical Fiber Communication Conference, San Diego, CA, USA, 11–15 March 2018.
58. Hadi, M.U.; Basit, A. Machine Learning for Performance Enhancement in Fronthaul Links for IOT Applications. In Proceedings of the 2021 International Conference on Digital Futures and Transformative Technologies (ICoDT2), Islamabad, Pakistan, 20–21 May 2021; pp. 1–5.
59. Liu, E.; Yu, Z.; Yin, C.; Xu, K. Nonlinear distortions compensation based on artificial neural networks in wideband and mul-ti-carrier systems. *IEEE J. Quantum Electron.* **2019**, *55*, 800305. [CrossRef]
60. Liu, J.; Zou, X.; Bai, W. Performance enhancement of UFMC based radio over fiber system using ANN equalizer. In Proceedings of the Asia Communications and Photonics Conference, Hangzhou, China, 26–29 October 2018.
61. Safari, L.; Baghersalimi, G.; Karami, A.; Kiani, A. On the Equalization of an OFDM-Based Radio-over-Fiber System Using Neural Networks. *Radioengineering* **2017**, *26*, 162–169. [CrossRef]
62. Zhou, Q.; Lu, F.; Xu, M.; Peng, P.-C.; Liu, S.; Shen, S.; Zhang, R.; Yao, S.; Finkelstein, J.; Chang, G.-K. Enhanced Multi-Level Signal Recovery in Mobile Fronthaul Network Using DNN Decoder. *IEEE Photonics Technol. Lett.* **2018**, *30*, 1511–1514. [CrossRef]
63. Xu, T.; Shevchenko, A.N.; Zhang, Y.; Jin, C.; Zhao, J.; Liu, T. Information rates in Kerr non-linearity limited optical fiber communication systems. *Opt. Express* **2021**, *29*, 17428–17439. [CrossRef] [PubMed]
64. Jin, C.; Shevchenko, N.A.; Li, Z.; Popov, S.; Chen, Y.; Xu, T. Nonlinear Coherent Optical Systems in the Presence of Equalization Enhanced Phase Noise. *J. Lightwave Technol.* **2021**, *39*, 4646–4653. [CrossRef]
65. Khayatzadeh, R.; Poëtte, J.; Cabon, B. Impact of Phase Noise in 60-GHz Radio-Over-Fiber Communication System Based on Passively Mode-Locked Laser. *J. Lightwave Technol.* **2014**, *32*, 3529–3535. [CrossRef]
66. Maeda, J.; Katoh, T.; Ebisawa, S. Effect of Fiber Dispersion on Subcarrier QAM Signal in Radio-Over-Fiber Transmission. *J. Lightwave Technol.* **2012**, *30*, 2625–2632. [CrossRef]
67. Hadi, M.U. Mitigation of nonlinearities in analog radio over fiber links using machine learning approach. *ICT Express* **2021**, *7*, 253–258. [CrossRef]

68. Vieira, L.C.; Gomes, N.J. Experimental demonstration of digital predistortion for orthogonal frequency-division multi-plexing-radio over fibre links near laser resonance. *IET Optoelectron.* **2015**, *9*, 310–316. [CrossRef]
69. Hadi, M.U.; Nanni, J.; Venard, O.; Baudoin, G.; Polleux, J.-L.; Traverso, P.A.; Tartarini, G. Linearity Improvement of VCSELs based Radio over Fiber Systems utilizing Digital Predistortion. *Adv. Sci. Technol. Eng. Syst. J.* **2019**, *4*, 156–163. [CrossRef]
70. Hadi, M.U.; Nanni, J.; Venard, O.; Baudoin, G.; Polleux, J.-L.; Tartarini, G. Practically Feasible Closed-Loop Digital Predistortion for VCSEL-MMF-Based Ra-dio-over-Fiber links. *Radioengineering* **2020**, *29*, 37–43. [CrossRef]
71. Hadi, M.U.; Basit, A.; Khurshid, K. Nonlinearities Mitigation in Radio over Fiber Links for Beyond 5G C-RAN Applications using Support Vector Machine Approach. In Proceedings of the 2020 IEEE 23rd International Multitopic Conference (INMIC), Bahawalpur, Pakistan, 5–7 November 2020; pp. 1–5.
72. Cao, W.; Li, Y.; Zhu, A. Magnitude-selective affine function based digital predistorter for RF power amplifiers in 5G small-cell transmitters. In Proceedings of the 2017 IEEE MTT-S International Microwave Symposium (IMS), Honololu, HI, USA, 4–9 June 2017; pp. 1539–1541.
73. Li, Y.; Cao, W.; Zhu, A. Instantaneous sample indexed magnitudes elective affine function-based behavioral model for digital predistortion of RF power amplifiers. *IEEE Trans. Microw. Theory Tech.* **2018**, *66*, 5000–5010.
74. LTE. *Evolved Universal Terrestrial Radio Access (E-UTRA)*; Base Station (BS) Radio Transmission and Reception; ETSI, TS 136 104 V8.2.0; European Telecommunications Standards Institute: Valbonne, France, 2018.
75. Xilinx DPD Design. Available online: https://www.xilinx.com/publications/prod_mktg/pn2061.pdf (accessed on 1 September 2021).
76. Hadi, M.U.; Traverso, P.A.; Tartarini, G.; Venard, O.; Baudoin, G.; Polleux, J.-L. Digital Predistortion for Linearity Improvement of VCSEL-SSMF-Based Radio-Over-Fiber Links. *IEEE Microw. Wirel. Compon. Lett.* **2019**, *29*, 155–157. [CrossRef]

applied sciences

MDPI

Article

Effect of Substrate Misorientation on the Structural and Optical Characteristics of In-Rich InGaAs/GaAsP Quantum Wells

Zhiwei Li [1,2], Yugang Zeng [1,2,*], Yue Song [1,2,*], Jianwei Zhang [1,2,*], Yinli Zhou [1,2], Yongqiang Ning [1,2], Li Qin [1,2] and Lijun Wang [1,2]

1 State Key Laboratory of Luminescence and Application, Changchun Institute of Optics, Fine Mechanics and Physics, Chinese Academy of Sciences, Changchun 130033, China; lizhiwei17@mails.ucas.ac.cn (Z.L.); zhouyinli@ciomp.ac.cn (Y.Z.); ningyq@ciomp.ac.cn (Y.N.); qinl@ciomp.ac.cn (L.Q.); wanglj@ciomp.ac.cn (L.W.)

2 Center of Materials Science and Optoelectronics Engineering, University of Chinese Academy of Sciences, Beijing 100049, China

* Correspondence: zengyg@ciomp.ac.cn (Y.Z.); songyue@ciomp.ac.cn (Y.S.); zjw1985@ciomp.ac.cn (J.Z.); Tel.: +86-138-4308-8351 (Y.Z.)

Abstract: InGaAs quantum well (QW) lasers have attracted significant attention owing to their considerable potential for applications in optical communications; however, the relationship between the misorientation of the substrates used to grow InGaAs QWs and the structural and optical properties of QWs is still ambiguous. In this study, In-rich InGaAs/GaAsP single QWs were grown in the same run via metal organic chemical vapor deposition on GaAs (001) substrates misoriented by 0°, 2°, and 15° toward (111). The effects of substrate misorientation on the crystal quality and structural properties of InGaAs/GaAsP were investigated by X-ray diffraction and Raman spectroscopy. The 0° substrate exhibited the least lattice relaxation, and with increasing misorientation, the degree of lattice relaxation increased. The optical properties of the InGaAs/GaAsP QWs were investigated using temperature-dependent photoluminescence. An abnormal S-shaped variation of the peak energy and inverse evolution of the spectral bandwidth were observed at low temperatures for the 2° substrate, caused by the localization potentials due to the In-rich clusters. Surface morphology observations revealed that the growth mode varied with different miscuts. Based on the experimental results obtained in this study, a mechanism elucidating the effect of substrate miscuts on the structural and optical properties of QWs was proposed and verified.

Keywords: misorientation; optical communication; InGaAs/GaAsP quantum well; optical properties; localization potential

Citation: Li, Z.; Zeng, Y.; Song, Y.; Zhang, J.; Zhou, Y.; Ning, Y.; Qin, L.; Wang, L. Effect of Substrate Misorientation on the Structural and Optical Characteristics of In-Rich InGaAs/GaAsP Quantum Wells. *Appl. Sci.* **2021**, *11*, 8639. https://doi.org/10.3390/app11188639

Academic Editors: Giuseppe Rizzelli and John Xiupu Zhang

Received: 9 August 2021
Accepted: 12 September 2021
Published: 17 September 2021

1. Introduction

Long wavelength semiconductor laser diodes have been an important technology in recent years, because of their major role in optical communications [1]. These lasers emit at wavelengths near the infrared band, and the InGaAs material system has shown promise as an active layer candidate for realizing wavelengths beyond 1 μm. Because the large lattice mismatch between InGaAs and GaAs substrates limits the layer thickness and heterointerface smoothness, several studies have been performed to improve the interface roughness and crystal quality. Nagle et al. improved the quality of the GaAs/InGaAs interface using short annealing steps at various growth temperatures [2]. Ma et al. demonstrated the possibility of avoiding strain accumulation in InGaAs layers using a gas-switching technique during growth [3]. In order to achieve a longer excitation wavelength, it is necessary to increase the indium (In) content in InGaAs QWs. However, a high content of In tends to cause a significant lattice mismatch between the wells and barriers. Lattice-mismatched heteroepitaxy is always accompanied by the generation of strain and dislocations, which degrade the device performance [4]. In order to reduce or eliminate strain accumulation and the formation of defects in QWs, tensile-strain GaAsP barrier layers have been employed to compensate for the compressive strain in the InGaAs QW layer. Compressive strain in the well and tensile strain in the barrier can offset each other during epitaxial growth, which reduces the formation of defects, such as dislocations [5]. Therefore, radiative recombination and surface roughness can be improved. However, the precise strain-balanced condition is not straightforward, and the mismatch strain is unavoidable. The growth of high quality InGaAs/GaAsP QWs is vital for realizing high performance devices. Choi et al. found that a reduction in the growth temperature yielded highly strained epitaxial layers with decent optical properties presumably owing to the reduced formation and propagation of dislocations [6]. Hou et al. studied the influence of different compositions of phosphorus in the barrier layer on the production of a high quality and smooth heterointerface of InGaAs/GaAsP QWs [7]. Dong et al. investigated the influence of growth temperature on interfacial quality of InGaAs/GaAsP multiple quantum wells [8]. Therefore, the optimization of crystal growth conditions is crucial to facilitate the production of high-performance devices.

Substrate misorientation is known to be crucial for developing high-efficiency solar cells and high-performance QW diode lasers [9]. However, the correlation between the structural and optical properties of InGaAs QWs and GaAs substrates with different misorientation angles has not yet been thoroughly investigated. Lin et al. investigated the influence of a tilted substrate on the photoluminescence, mismatch, doping, thickness, and morphology of GaInP QWs and found that the 15° substrate was a suitable choice for use in a red light waveband laser diode wafer [10]. Botha et al. studied the influence of growth interruption and substrate misorientation on the optical and heterointerface properties; the use of the growth interruption technique during growth resulted in the 2° substrate being preferred for the growth of InGaAs/GaAs QWs [11]. They mainly explored the effect of epitaxial growth conditions on the structure of materials with little lattice mismatch on misoriented substrate. However, there are few studies about the effect of substrate misorientation on QWs with large lattice mismatch. The use of vicinal substrates for growing InGaAs/GaAsP QWs has become technologically significant because they result in improved optical properties [9]. However, the correlation between the structural and optical properties of In-rich InGaAs QWs and GaAs substrates with different misorientation angles has not been elucidated.

In this study, the dependence of structural and optical properties of large-mismatch InGaAs/GaAsP QWs on substrate misorientation were investigated and analyzed in detail. This provides a better experimental method and theoretical basis for the preparation of novel optoelectronic materials and devices.

2. Materials and Methods

In-rich InGaAs/GaAsP QWs were grown on n-type GaAs (001) substrates misoriented by 0°, 2°, and 15° toward (111) using a MOCVD AIXTRON 200/4 reactor under identical growth conditions. The following reactants were used: trimethylindium (TMIn), trimethylgallium (TMGa), arsine (AsH_3), and phosphine (PH_3). Purified H_2 was used as a carrier gas. The growth pressure and temperature of the MOCVD system were set to 100 mbar and 630 °C, respectively, a growth rate of 0.5 nm/s during the growth and V/III ratios of 80 were employed. The structure of an InGaAs/GaAsP QW is depicted in Figure 1. For all three samples, the substrate surface was deoxidized via thermal desorption at 680 °C for 5 min under an AsH_3 atmosphere. A 300 nm GaAs buffer layer was subsequently grown on the 2-inch oriented GaAs substrate using TMGa and AsH_3 with a growth time of ~600 s. A 5 nm GaAsP barrier was subsequently grown, followed by the growth of a 7 nm InGaAs QW. Finally, a 5 nm GaAsP barrier with a PH_3:AsH_3 ratio of 3.5:1 in the vapor phase was employed as the capping layer.

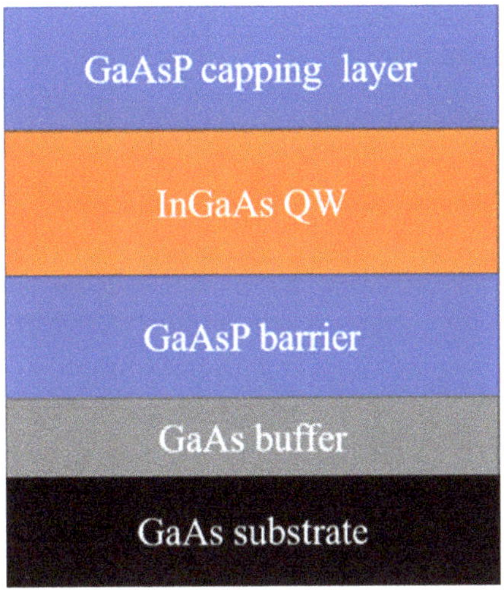

Figure 1. Schematic of the InGaAs/GaAsP quantum well.

To achieve lasing at near-infrared wavelength in this study, the In and phosphorous compositions were fixed at 38% and 20% in the InGaAs QWs and GaAsP barriers, respectively; the thicknesses of the QWs and barriers were estimated to be 7 and 5 nm, respectively. Lattice constants are known to change with composition, and the effects of the composition of In on lattice constants can be expressed as follows:

$$a_{\mathrm{In}_x\mathrm{GaAs}} = a_{\mathrm{GaAs}} + (a_{\mathrm{InAs}} - a_{\mathrm{GaAs}}) \cdot x. \quad (1)$$

The strain in the layers can be defined as follows:

$$\varepsilon = \frac{\Delta a}{a_{sub}} \quad (2)$$

where Δa is the difference between the lattice constants of the two materials and a_{sub} is the lattice constant of the substrate. The lattice constant of $In_{0.38}GaAs$ is obtained as 5.807 Å using Equation (1) and the compressive strain of the $In_{0.38}Ga_{0.62}As$ layer is evaluated to be 27,224 ppm (Equation (2)). Tensile strain GaAsP barrier layers were employed to compensate for the compressive strain in the InGaAs QW layer, and then the defect density was reduced. The average strain is defined as [7]

$$(T_{\mathrm{InGaAs}} \times \varepsilon_{\mathrm{InGaAs}} + T_{\mathrm{GaAs}P} \times \varepsilon_{\mathrm{GaAs}P})/(T_{\mathrm{InGaAs}} + T_{\mathrm{GaAs}P}). \quad (3)$$

where T and ε represent the thickness and strain of each layer, respectively. ε is positive if the layer has compressive strain and is negative for tensile strain. The accumulated strain in the InGaAs/GaAsP QW was 7006 ppm, which indicated that it was not fully compensated and residual strain was likely present in the material.

To investigate the crystal quality, and strain mismatch of the epitaxial layers, the samples were identified by X-ray diffraction (XRD, Bruker D8). A Raman spectrometer (RENISHAW InVia) was employed to study the structural disorder and residual strain in the epitaxial layers of the three samples. Atomic force microscopy (AFM, Bruker MultiMode 8) was used to conduct morphological analysis and determine the RMS surface roughness. The optical characteristics of the InGaAs/GaAsP QW were investigated using temperature-dependent photoluminescence spectroscopy (ARS 8200 cryogenic system) with a 488 nm continuous-wave laser.

3. Results and Discussion

Figure 2 shows the XRD patterns of the InGaAs/GaAsP QWs with different miscut angles. The strongest peak originated from the GaAs substrates, and the satellite peaks of the QW are clearly visible in samples with miscut angles of 0° and 2°. However, the satellite peaks of the InGaAs/GaAsP QW disappeared in the XRD pattern of the 15° sample, which indicated its poor interfacial quality [12]. From the XRD patterns of the three samples in Figure 2, the relax are obtained by the simulation of the diffraction patterns, 5.03% strain relaxation occurs in the 0° sample while 19.26% strain relaxes in the 2° sample, as for the 15° sample, 90.58% strain relaxation occurs [13]. The XRD result show that the crystal quality degraded with increasing the angle of substrate misorientation.

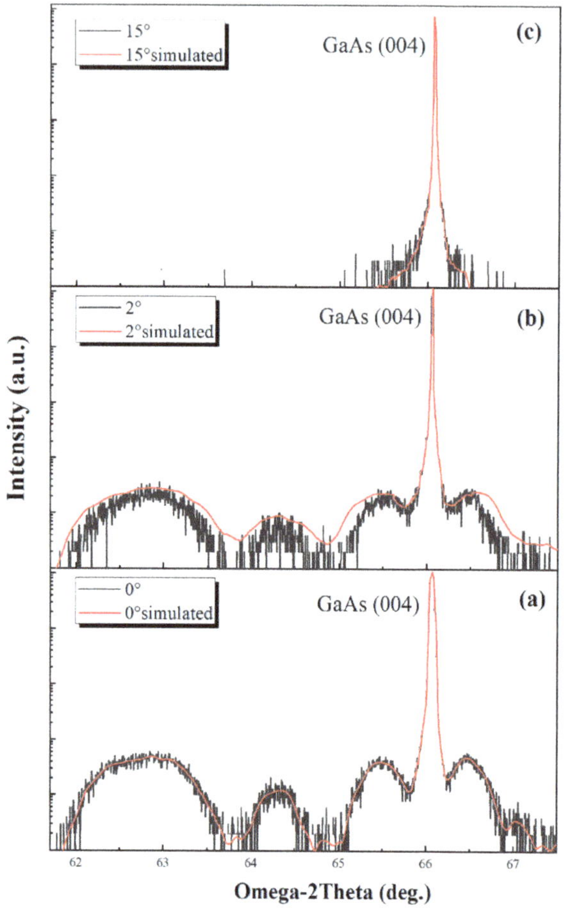

Figure 2. Omega-2θ scans around the (004) symmetric XRD patterns of InGaAs/GaAsP QWs grown on substrates with misorientations of (**a**) 0°, (**b**) 2°, and (**c**) 15°.

Raman scattering is an important technique used to investigate the optical properties of materials, wherein the disorder and local strain phenomena in materials can be adequately represented by vibrational spectra. The disorder and strain in materials can alter the phonon frequency and phonon lifetime, which can lead to the broadening of the Raman spectrum and breaking of Raman selection rules [14]. The frequency shifts of longitudinal optical (LO) phonons are extremely sensitive to strain; compressive and tensile strain are known to result in positive and negative frequency shifts, respectively [15].

The samples prepared in this study feature a lattice mismatch between the In-rich InGaAs QW and the GAs substrate. The In-rich layer generates a lattice mismatch stress in the adjacent epitaxial layer. The residual stress in the epitaxial layer can be obtained by measuring the frequency shift of the characteristic peak of GaAs via Raman scattering. The stress in the InGaAs material can be evaluated using the frequency shift of the GaAs-like LO mode of the InGaAs material; Equation (4) was utilized to calculate the frequency of the GaAs-like LO mode in the InGaAs epitaxial layer [16]:

$$\omega_{LO} = 290 - 18.6x - 32.5x^2, \tag{4}$$

where x is the content of In and ω_{LO} is the GaAs-like LO frequency in the strain-free InGaAs material as a function of x. The GaAs-like LO frequency in an ideal strain-free $In_{0.38}Ga_{0.62}As$ layer is evaluated to be 278.2 cm^{-1}. The residual strain in the epitaxial layer is evaluated using the deviation from the measured frequency of the GaAs-like LO mode to that of the ideal strain-free $In_{0.38}Ga_{0.62}As$ layer [17].

GaAs-related LO phonons from InGaAs/GaAsP QWs are known to consist of two LO phonon modes [18]—one from the GaAs buffer layer because of the thin QW that can be penetrated into the GaAs buffer layer [19], and the other from the InGaAs layer that exhibits a GaAs-like LO phonon peak. As shown by the dashed line in Figure 3, the GaAs-like LO frequency in the ideal strain-free $In_{0.38}Ga_{0.62}As$ epitaxial layer was obtained as 278.2 cm^{-1}. The GaAs-like LO frequency of the 2° GaAs substrate was observed at 283.4 cm^{-1} via fitting of the Raman spectra; this sample features the smallest shift with respect to 278.2 cm^{-1} compared to that of the 0° GaAs substrate (285.2 cm^{-1}), which indicated that the strain induced via lattice mismatch was partially relaxed in the epitaxial layer in the 2° GaAs substrate. The GaAs-like LO frequency of the 15° GaAs substrate was observed at 279.8 cm^{-1} which is very close to the ideal strain-free $In_{0.38}Ga_{0.62}As$ layer and indicated that substantial strain had relaxed in 15° substrate.

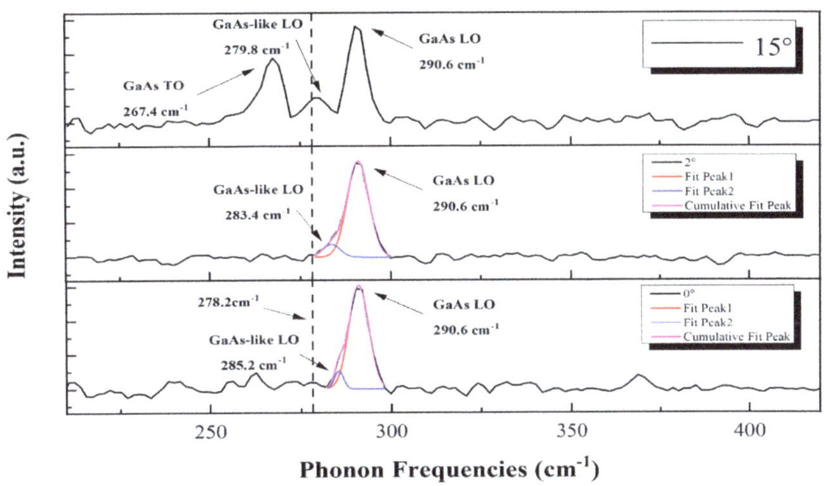

Figure 3. Raman spectra of InGaAs/GaAsP QWs grown on a GaAs (001) substrate with misorientations of 0°, 2°, and 15° using a 488 nm laser.

The full width at half maximum (FWHM) values of the GaAs-like LO frequencies for the 0° and 2° substrates were obtained as 3.0 and 5.6 cm^{-1}, respectively, by fitting the Raman spectra. The increase in the FWHM of GaAs-like LO phonons suggests that the residual strains released through defect generation and structural disorder [15,20]. Consequently, the epitaxial layer on the 2° GaAs substrate exhibited a poorer crystalline quality than that of the 0° GaAs substrate.

Raman selection rules are often applied for analyzing the crystal quality; the LO phonon is allowed and the TO phonon is forbidden for the (001) plane of a sphalerite crystal structure [21–24]. The disorder in InGaAs films can be assessed by the ratio between the TO and LO peak intensities because the presence of the TO phonon is an indicator of crystal structural disorder. As shown in Figure 3, the TO mode at 267.4 cm^{-1} appeared only in the 15° sample, confirming its extremely large degree of crystal disorder compared to that of the other two samples [25]. The above results show that the strain relaxes gradually with the increase of the misorientation. These results follow trends similar to those observed in the XRD results.

AFM analysis was subsequently conducted to verify the growth model of the InGaAs/GaAsP QWs. Figure 4 shows the AFM results on the surface morphologies of the three samples with different misorientations. Figure 4a reveals that the 0° sample exhibited a step-flow characteristic with RMS surface roughness of 0.147 nm, which indicated the high quality of the heterointerface in the 0° sample. Figure 4b indicated the narrowing of the terrace width in the 2° sample; the RMS surface roughness increased from 0.147 to 0.363 nm due to the composition-related inhomogeneity of In clusters that lead to a partial internal strain relaxation. In this case, the localization potential was formed because of the In clusters. Figure 4c reveals the presence of self-assembled quasi-quantum dots in the 15° sample. Randomly positioned growth nuclei were formed and the local strain subsequently relaxed at the locations of these nuclei [26,27]; a surface roughness of 0.479 nm was obtained. Based on these results, a growth model is proposed to elucidate the different morphologies with the substrates featuring different misorientations.

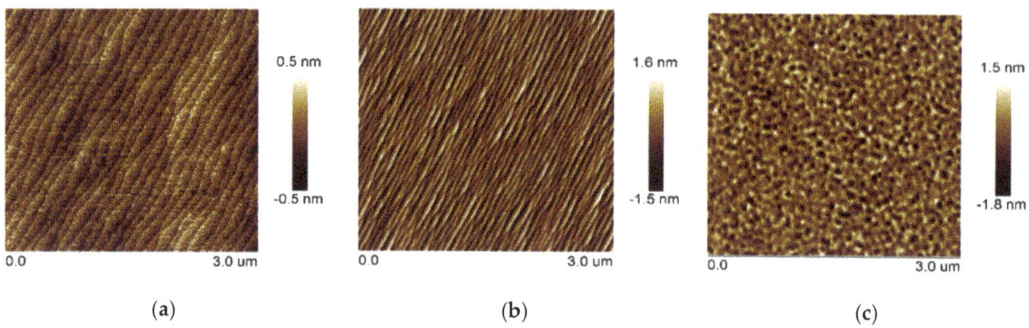

Figure 4. AFM images of InGaAs/GaAsP QWs in a 3 × 3 μm^2 scan area with different misoriented substrates: (**a**) 0°, (**b**) 2°, and (**c**) 15°.

Figure 5 illustrates a surface on a substrate that is composed of steps and terraces. The growth condition in this scenario is presumed to involve surface coverage of the misoriented substrate with As atoms, because of the high V/III ratio (~150). Two dangling bonds are present on the terrace and three dangling bonds are at the kinks of the terrace edges. The incorporation probability of group III atoms into step kinks is presumed to be higher than that on the terraces. Upon reaching the growing surface, the group III species preferentially adhere to the step and kink sites on the terrace edge. This results in a step flow growth mode. The width of the terrace is known to decrease with increasing substrate misorientation, owing to the high density of growth steps on the tilted substrates [28].

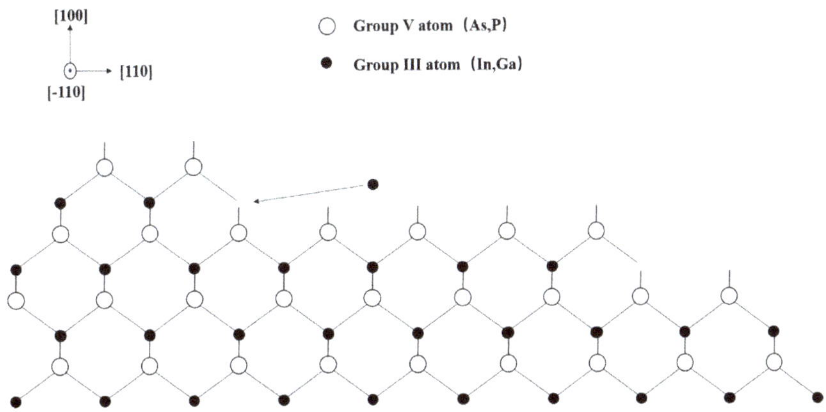

Figure 5. Schematic illustration of the growth process.

The disparity in the samples from the 0° to the 2° substrate was possibly caused by the inequality of diffusion lengths of the In and Ga cations for the growth of InGaAs. Ga cations possess smaller diffusion lengths than those of In; therefore, In atoms are likely to be mobile enough to incorporate at the step edges of the 2° substrate. The bonds in the kink sites are occupied with In cations, and therefore, Ga atoms tend to incorporate at the flat terrace [29]. If this occurs, a severe compositional nonuniformity can arise owing to a "step-induced" clustering of In atoms, which can lead to nonuniform components and the formation of local energy level states, and partial strain relaxation occurs with the formation of in clusters. This modulation, or clustering, did not occur in the 0° substrate due to the absence of a high density of step edges. As for 15° sample the formation of randomly positioned growth nuclei and the subsequent local strain relaxation at the nuclear locations could occur owing to the large degree of misorientation, resulting in high strain relaxation and the formation of island.

To study the optical characteristics of the InGaAs/GaAsP QWs on different substrate orientations, temperature-dependent photoluminescence analysis was performed. Figure 6 shows the temperature-dependent PL spectra of three samples. For the 0° and 15° samples, it can be seen that the PL intensity decreased and the PL peak wavelength had a red-shift with increasing temperature. But the peak position of the 15° sample reached 1206 nm at room temperature and its extremely high FWHM and relatively low intensity indicate the PL characteristics of self-assembled quantum dots. The results are consistent with the AFM results shown in Figure 4c. However, the 2° sample exhibited an anomalous blue shift tendency of PL peak wavelength in the temperature range from 40 K to 85 K; the reason is discussed in detail below.

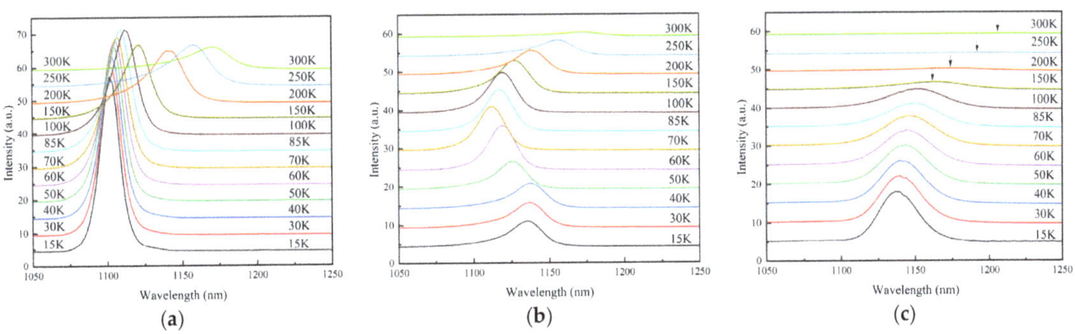

Figure 6. Temperature-dependent of PL spectra for InGaAs/GaAsP QWs on (**a**) 0°, (**b**) 2°, and (**c**) 15° misoriented substrates.

Figure 7 shows the temperature dependence of the PL peak energy of the three samples. The PL peak position of the 0° sample shifted toward lower energies as the temperature increased over the entire range; this typical behavior can be well fitted using the Varshni empirical formula [30]:

$$E_g(T) = E_g(0) - \alpha T^2/(\beta + T), \quad (5)$$

where $E_g(0)$ is the energy gap between nominal subband edges at 0 K, and α and β are the Varshni equation parameters. The optimizing parameters, α and β, obtained from optimal fits of the QW data are listed in Table 1.

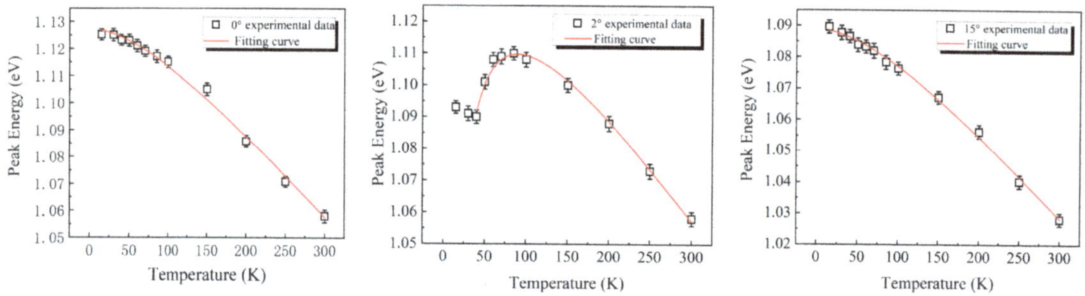

(a) (b) (c)

Figure 7. Temperature-dependent peak energies of the InGaAs/GaAsP QWs on (**a**) 0°, (**b**) 2°, and (**c**) 15°-misoriented substrates.

Table 1. Fitting parameters, α, β, and σ, for the differently misoriented substrates.

Sample	$E_g(0)$ (eV)	α (eV/K)	β (K)	σ (meV)
0°	1.127	3.49×10^{-4}	150	
2°	1.153	4.03×10^{-4}	109	14.7
15°	1.089	2.84×10^{-4}	120	

The peak energies of systems with the 0° and 15° substrates can be well fitted using Equation (5); however, the data corresponding to the 2° sample cannot be fitted using the Varshni empirical formula in the low-temperature range (T < 60 K). An anomalous band gap evolution in the form of an S-shaped (redshift–blueshift–redshift) PL peak energy was observed at low temperatures. This abnormal behavior may be attributed to the localization effect caused by composition fluctuation [31–35]. Eliseev et al. amended the Varshni empirical equation using band-tail states to fit this type of experimental data as follows [36]:

$$E_g(T) = E_g(0) - \alpha T^2/(\beta + T) - \sigma^2/k_B T, \quad (6)$$

where k_B is the Boltzmann constant and σ represents the degree of the localization effect. The fitting parameters are listed in Table 1.

Figure 7b indicates that the data representing the 2° substrate can be adequately fitted in the 60–300 K temperature region. However, the aforementioned empirical models cannot accurately describe the temperature dependence of the peak energy in the InGaAs/GaAsP QW with a 2° substrate at temperatures below 60 K; the PL peak of the 2° substrate exhibited a redshift as the temperature increased from 15 to 40 K. The S-shaped variation over the entire temperature range occurred because of the existence of different localization potentials [37,38]. A possible explanation for the S-shaped behavior of the peak energy is illustrated in Figure 8.

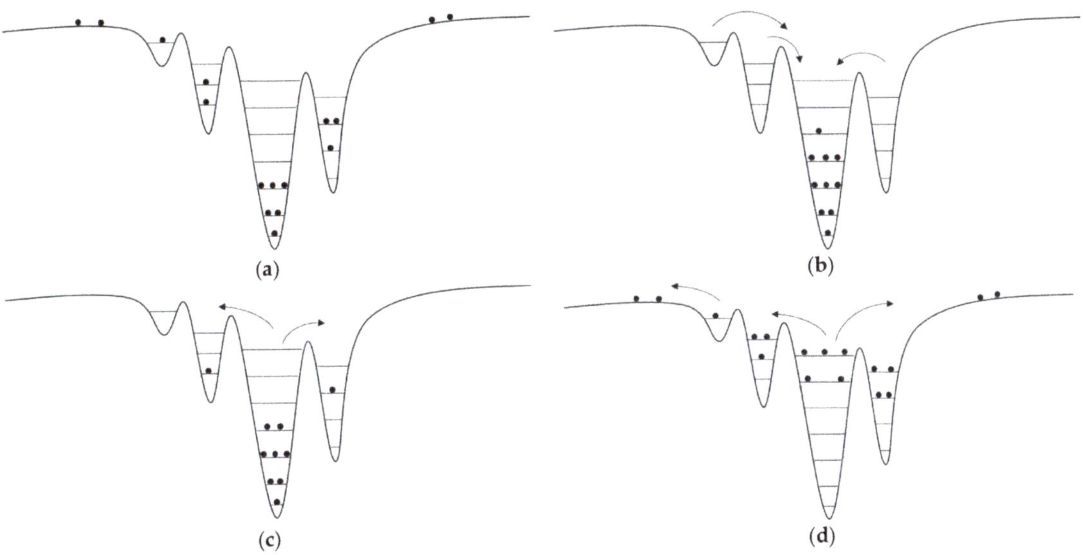

Figure 8. Schematic diagrams illustrating the possible mechanism of the S-shaped temperature-dependent PL peak energy. (**a**) At the lowest temperature, carriers exist in all possible potential minimums. (**b**) At a slightly higher temperature, carriers relax to the deepest potentials. (**c**) As the temperature continues to rise, carriers redistribute to higher energies. (**d**) As the temperature rises very high, most carriers escape from the localization potentials.

At low temperatures, carriers are freely distributed throughout the QW layer (Figure 8a). With increasing temperatures, the weakly localized carriers are thermally activated and have a greater chance to relax to the deep localized states (Figure 8b); this results in a red shift in the PL peak energy. A further increase in the temperature leads to the carriers in the deep localized states obtaining sufficient energy to repopulate the shallow localized states (Figure 8c), which produces a blueshift in the emission peak energy. As the temperature continues to rise, the thermal carriers escape from the localized states into the barrier states and become free carriers (Figure 8d); at even higher temperatures, the peak energy exhibited redshifts owing to the shrinkage of the temperature-induced band gap upon thermal activation and delocalization of the carriers [39].

Typically, two factors are known to affect the PL linewidth of the InGaAs/GaAsP QWs, one is inhomogeneous broadening which is primarily determined by fluctuations of circumstance, for example, interface roughness, composition variation, and crystal disorder. The other one is homogeneous which is closely related to temperature and is primarily affected by the interaction between electrons and phonons [40,41].

Figure 9 shows the behavior of the FWHM with respect to temperature. Inhomogeneous broadening and exciton interactions with phonons are considered. The trends of FWHM conform to the following expression [42]:

$$\Gamma(T) = \Gamma_0 + aT + \gamma/[(\exp(E_{LO}/k_B T) - 1], \tag{7}$$

where $\Gamma(T)$ represents the FWHM of the photoluminescence spectra, Γ_0 represents the inhomogeneous contribution to the FWHM that is related to the interface roughness and crystal disorder, a is a parameter representing the acoustic phonon scattering strength, γ is a parameter that is related to the exciton–phonon coupling, and E_{LO} is the longitudinal optical phonon energy of QW layer. The experimental data of the three samples were fitted using Equation (7) and shown in Figure 9; the fitted parameters are listed in Table 2. The value of Γ_0 is noted to increase with increasing misorientation, which indicates that the interface roughness and alloy disorder increased as the substrate misorientation increased.

A higher value of E_{LO} is known to indicate a larger strain in the epitaxial layer [13,42]. The highest E_{LO} value corresponded to the 0° sample and indicated the largest strain in this sample compared to those of the others; considerable strain relaxation was noted to occur in the 2° and 15° samples, which had also been implied by the Raman scattering results.

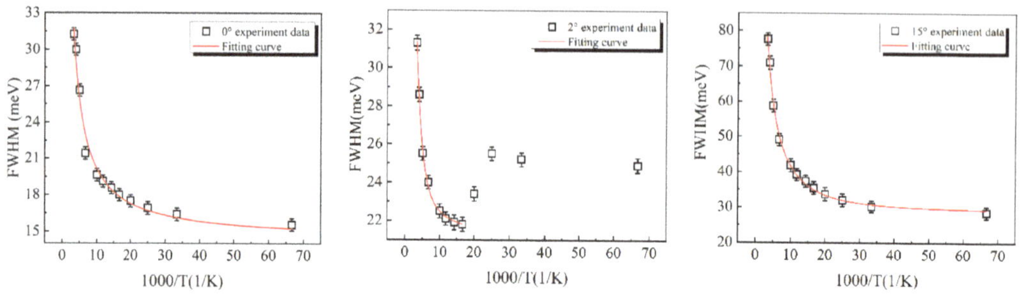

Figure 9. PL full width at half maximum (FWHM) as a function of 1000/T of the InGaAs/GaAsP QWs on substrates misoriented by (**a**) 0°, (**b**) 2°, and (**c**) 15°. The fitting curves of the FWHM data of the PL spectra are predictions based on the exciton–optical phonon coupling model.

Table 2. FWHM-fitting parameters for samples with different misoriented substrates.

Sample	Γ_0 (meV)	a (meV/K)	γ (meV)	E_{LO} (meV)
0°	14.36	0.058	1	0.83
2°	20.91	0.015	43	0.05
15°	28.00	0.081	12	0.01

The evolution of FWHM vs. temperature revealed a unique tendency before 70 K for the 2° sample. The FWHM increased slowly below 40 K, sharply decreased in the 40–70 K range, and increased above 70 K. The FWHM trends exhibited a reverse temperature-dependent relationship compared to that of the PL peak energy. The FWHM data for the 2° sample are plotted against the corresponding peak energies in Figure 10, with the temperature spanning the entire region from 15 to 300 K. The fitting linear curve was obtained by the least square method drawn by the solid red line, a clearly negative linear relationship can be observed between them, which indicates that the evolutions of peak energy and FWHM of QW with temperature obey a similar or associated mechanism [43].

At low temperature (<40 K), carriers are thermally activated upon an increase in the temperature and have a significant opportunity to transit to deeper localization potentials; therefore, the range of energy levels of the carrier distribution widens, which leads to an increment in the FWHM. As the temperature increases to a critical level, the carriers obtain sufficient energy to enable their delocalization. The decrease in FWHM in the temperature range of 40–70 K was due to the process of carrier escape from the located states. The increment in FWHM at temperatures above 70 K presumably occurred because of a combination of the thermalization of carriers and phonon scattering caused by the increase in temperature. It is worth noting that the variations of PL peak position and FWHM of the 15° sample were not similar to the 2° sample. In addition, the value of the PL peak and FWHM of the 15° sample were much higher than the 0° and 2° samples, this is because large misorientations yield substantial strain relief and random quantum dots can form, therefore the luminescence characteristics are dominated by quantum dots.

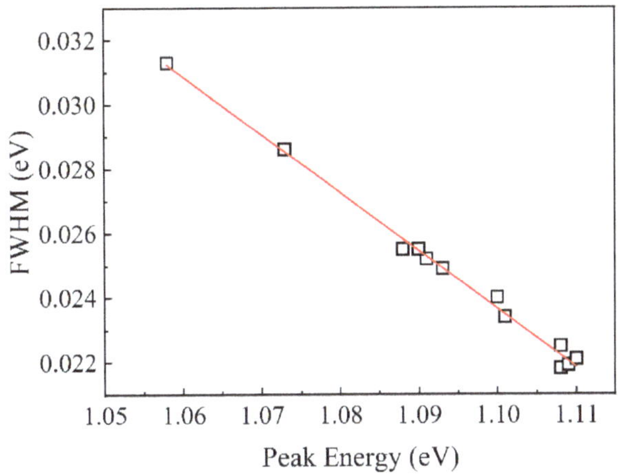

Figure 10. Relationship between photon energy and FWHM for the 2° sample. The red solid line represents the curve fit to the presented data.

The temperature-dependent PL intensity data can be fit using the Arrhenius equation shown below to study the mechanism of PL quenching with increasing temperatures [44]:

$$I(T) \propto 1/(1 + \sum_i C_i \exp(-E_i/k_B T)). \tag{8}$$

$I(T)$ represents the integrated PL intensity at temperature T, C_i represent the factors related to the densities of nonradiative recombination centers, and E_i represent the activation energies of the relevant nonradiative recombination centers. Two types of nonradiative recombination channels have been proposed for the quenching mechanism of PL intensities in epitaxially grown materials. The modified Arrhenius equation used to fit these experimental data can be expressed as [45–47]

$$I(T) \propto 1/(1 + C_1 \exp(-E_A/k_B T) + C_2 \exp(-E_B/k_B T)), \tag{9}$$

where E_A and E_B are the activation energies of the corresponding nonradiative recombination processes, and C_1 and C_2 are two constants that are related to the number of nonradiative recombination centers in the QWs. The modified Arrhenius equation provides excellent fits for the integrated PL intensity data corresponding to the 0° and 15° samples and for data obtained at temperatures above 60 K for the 2° sample, as shown by the red lines in Figure 11. The parameters used to obtain the optimal fits are listed in Table 3.

Two different nonradiative recombination processes have been typically described. The first process involves thermally activated recombination in traps that are related to the incorporation of impurities [48,49], whereas the second involves the thermal emission of the carriers out of the located states with an activation energy related to the depth of the located states potential, followed by effective nonradiative recombination [38].

The fitting results revealed that all investigated samples (0°, 2°, and 15°) possessed nearly similar values of E_A and C_1. This suggests that the types of impurities in the three samples were similar because of them being grown under identical growth conditions. However, the E_B of the 2° sample was significantly higher than those of the 0° and 15° samples, because of the deep localized potential formed by the In-rich cluster during its epitaxial growth. The E_B of the 15° sample was lower than those of the other samples, because the quantum effect of quasi-nanocrystals caused by the severe relaxation of stress during the growth of the 15° sample and the diffusion of In atoms from the well to the barrier in nanodots reduce the potential barrier [50]. The high C_2 values of the 2° and

15° samples indicate the high defect density that was caused by stress relaxation during the growth process.

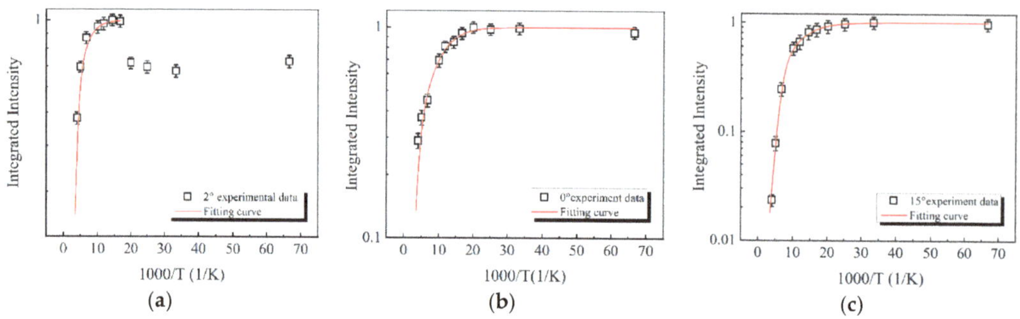

Figure 11. Temperature dependence of PL integrated intensities of the InGaAs/GaAsP QWs on substrates misoriented by (**a**) 0°, (**b**) 2°, and (**c**) 15°; the data are fitted using the Arrhenius model. An abnormal temperature-dependent PL intensity is observed in the sample with the 2°-misoriented substrate.

Table 3. Parameters obtained by fitting the Arrhenius model for data representing samples with differently misoriented substrates.

Sample	C_1	E_A (meV)	C_2	E_B (meV)
0°	5.30	21.72	260	106.55
2°	4.01	35.25	1100	148.47
15°	5.96	18.72	1525	87.26

4. Conclusions

The effect of substrate misorientation on the structure and optical properties of In-rich InGaAs/GaAs QW structures grown via MOCVD was investigated. Results obtained via XRD and Raman spectroscopy indicated that the strain in the InGaAs/GaAsP decreased monotonically with increasing misorientation. AFM was used to probe the surface morphology and microstructures of the samples, which revealed that with a substrate misorientation of 0°, the growth mode became a step-flow mode and the heterointerface smoothness was degraded as the substrate misorientation increased. The growth mode transformed into three-dimensional growth as the substrate misorientation increased to 15°. Based on these results, the growth of the InGaAs/GaAsP QW nanostructures was modeled from the viewpoint of atomic step motion; a mechanism associated with the accumulation of In on tilted substrates was proposed and verified. The localization potentials formed by the In clusters were primarily caused by the appropriate misoriented substrates and the preferential of In/Ga atoms on the terrace to the steps. Based on these results and the parameters related to the temperature-dependent photoluminescence of the samples, the strain relaxation degree of 0° substrate is the minimum due to the absence of a high density of step edges, a mechanism was proposed for the S-shaped temperature dependence of the peak energy and FWHM of the 2° sample. Partial strain relaxation occurs with the formation of in clusters owing to the relatively narrow steps, when the substrate misorientation increases to 15°, the formation of randomly positioned growth nuclei yield substantial strain relaxation.

Author Contributions: Conceptualization, Z.L. and Y.Z. (Yugang Zeng); methodology, Z.L. and Y.Z. (Yinli Zhou); software, Z.L.; validation, Y.N., Y.Z. (Yugang Zeng) and Y.S.; formal analysis, J.Z.; investigation, Z.L.; resources, Y.N.; data curation, Y.Z. (Yugang Zeng); writing—original draft preparation, Z.L.; writing—review and editing, Y.Z. (Yugang Zeng) and Y.S.; visualization, Y.S. and L.Q.; supervision, Y.N.; project administration, Y.N. and L.W.; funding acquisition, Y.S. and L.Q. All authors have read and agreed to the published version of the manuscript.

Funding: This research was funded by National Natural Science Foundation of China (NSFC) (61904179, 62090060, 61727822, 61991433, 61604151, 11874353, 61935009, 61934003, 61674148, 61805236); National Science and Technology Major Project of China (2018YFB2200300, 2016YFE0200700, 2017YFB0503100, 2017YFB0405100, 2018YFB0504600); CAS Projection of instrument development (YJKYYQ20180045); Frontier Science Key Program of the President of the Chinese Academy of Sciences (QYZDY-SSW-JSC006); Key R & D projects in Guangdong Province (2020B090922003); Equipment pre research (41414060103); Pilot project of CAS (XDB43030302); Science and Technology Development Project of Jilin Province (20200401069GX, 20200401062GX, 20190302042GX).

Institutional Review Board Statement: Not applicable.

Informed Consent Statement: Not applicable.

Data Availability Statement: The data presented in this study are available on request from the corresponding author.

Conflicts of Interest: The authors declare no conflict of interest.

References

1. Hou, C.-C.; Chen, H.-M.; Zhang, J.-C.; Zhuo, N.; Huang, Y.-Q.; Hogg, R.-A.; Childs, D.T.D.; Ning, J.-Q.; Wang, Z.-G.; Liu, F.-Q.; et al. Near-infrared and mid-infrared semiconductor broadband light emitters. *Light-Sci. Appl.* **2018**, *7*, 7. [CrossRef] [PubMed]
2. Nagle, J.; Landesman, J.P.; Larive, M.; Mottet, C.; Bois, P. Indium surface segregation in strained GaInAs quantum wells grown on GaAs by MBE. *J. Cryst. Growth* **1993**, *127*, 550–554. [CrossRef]
3. Ma, S.; Wang, Y.; Sodabanlu, H.; Watanabe, K.; Sugiyama, M.; Nakano, Y. Effect of hetero-interfaces on in situ wafer curvature behavior in InGaAs/GaAsP strain-balanced MQW. *J. Cryst. Growth* **2012**, *352*, 245–248. [CrossRef]
4. Tansu, N.; Mawst, L.J. High-performance strain-compensated InGaAs-GaAsP-GaAs (λ = 1.17 μm) quantum well diode lasers. *IEEE Photonics Technol. Lett.* **2001**, *13*, 179–181. [CrossRef]
5. Pfluegl, C.; Diehl, L.; Tsekoun, A.; Go, R.; Patel, C.K.N.; Wang, X.; Fan, J.; Tanbun-Ek, I.; Capasso, F. Room-temperature continuous-wave operation of long wavelength (λ = 9.5 μm) MOVPE-grown quantum cascade lasers. *Electron. Lett.* **2007**, *43*, 1026–1028. [CrossRef]
6. Choi, W.J.; Dapkus, D.; Jewell, J.J. 1.2-μm GaAsP/InGaAs strain compensated single-quantum-well diode laser on GaAs using metal-organic chemical vapor deposition. *IEEE Photonics Technol. Lett.* **1999**, *11*, 1572–1574. [CrossRef]
7. Hou, C.; Zou, Y.; Wang, H.; Wang, X.; Xu, Y.; Wang, Q.; He, Z.; Fan, J.; Shi, L.; Xu, L.; et al. Tailoring strain and lattice relaxation characteristics in InGaAs/GaAsP multiple quantum wells structure with phosphorus doping engineering. *J. Alloys Compd.* **2019**, *770*, 517–522. [CrossRef]
8. Dong, H.; Sun, J.; Ma, S.; Liang, J.; Xu, B. Investigation of the growth temperature on indium diffusion in InGaAs/GaAsP multiple quantum wells and photoelectric properties. *RSC Adv.* **2015**, *5*, 75211–75217. [CrossRef]
9. Sodabanlu, H.; Wang, Y.; Watanabe, K.; Sugiyama, M.; Nakano, Y. Growth of InGaAs/GaAsP multiple quantum well solar cells on mis-orientated GaAs substrates. *J. Appl. Phys.* **2014**, *115*, 233104. [CrossRef]
10. Lin, T.; Lin, N.; Xiong, C.; Zhong, L.; Qi, Q.; Zhao, Y.; Wang, C.; Guo, E.; Liu, S.; Ma, X. Effects of the substrate misorientation on the structural and optoelectronic characteristics of tensile GaInP quantum well laser diode wafer. *J. Alloys Compd.* **2016**, *688*, 83–87. [CrossRef]
11. Botha, J.R.; Leitch, A.W.R. Influence of growth interruption on the heterointerface morphology ofInGaAsGaAs strained quantum wells. *J. Cryst. Growth* **1996**, *169*, 629–636. [CrossRef]
12. Kwon, Y.H.; Gainer, G.H.; Bidnyk, S.; Cho, Y.H.; Song, J.J.; Hansen, M.; DenBaars, S.P. Structural and optical characteristics of multiple quantum wells with different In compositions. *Appl. Phys. Lett.* **1999**, *75*, 2545–2547. [CrossRef]
13. Dong, H.; Sun, J.; Ma, S.; Liang, J.; Lu, T.; Liu, X.; Xu, B. Influence of substrate misorientation on the photoluminescence and structural properties of InGaAs/GaAsP multiple quantum wells. *Nanoscale* **2016**, *8*, 6043–6056. [CrossRef] [PubMed]
14. Loudon, R. The Raman effect in crystals. *Adv. Phys.* **1964**, *13*, 423–482. [CrossRef]
15. Groenen, J.; Landa, G.; Carles, R.; Pizani, P.S.; Gendry, M. Tensile and compressive strain relief in epilayers grown on InP probed by Raman scattering. *J. Appl. Phys.* **1997**, *82*, 803–809. [CrossRef]
16. Emura, S.; Gonda, S.; Matsui, Y.; Hayashi, H. Internal-stress effects on Raman spectra of $In_xGa_{1-x}As$ on InP. *Phys. Rev. B Condens. Matter* **1988**, *38*, 3280–3286. [CrossRef]
17. Li, J.; Miao, G.; Zhang, Z.; Zeng, Y. Experiments and analysis of the two-step growth of InGaAs on GaAs substrate. *Crystengcomm* **2015**, *17*, 5808–5813. [CrossRef]
18. Dong, H.; Sun, J.; Ma, S.; Liang, J.; Lu, T.; Jia, Z.; Liu, X.; Xu, B. Effect of potential barrier height on the carrier transport in InGaAs/GaAsP multi-quantum wells and photoelectric properties of laser diode. *PCCP* **2016**, *18*, 6901–6912. [CrossRef]
19. Pistol, M.; Liu, X. Quantum-well structures of direct-band-gap $GaAs_{1-x}P_x$/GaAs studied by photoluminescence and Raman spectroscopy. *Phys. Rev. B Condens. Matter* **1992**, *45*, 4312–4319. [CrossRef]
20. Pizani, P.S.; Lanciotti, F.; Jasinevicius, R.G.; Duduch, J.G.; Porto, A.V.P. Raman characterization of structural disorder and residual strains in micromachined GaAs. *J. Appl. Phys.* **2000**, *87*, 1280–1283. [CrossRef]

21. Feng, Z.C.; Lin, H.C.; Zhao, J.; Yang, T.R.; Ferguson, I. Surface and optical properties of AlGaInP films grown on GaAs by metalorganic chemical vapor deposition. *Thin Solid Films* **2006**, *498*, 167–173. [CrossRef]
22. Feng, Z.C.; Armour, E.; Ferguson, I.; Stall, R.A.; Holden, T.; Malikova, L.; Wan, J.Z.; Pollak, F.H.; Pavlosky, M. Nondestructive assessment of $In_{0.48}(Ga_{1-x}Al_x)$ 0.52 P films grown on GaAs (001) by low pressure metalorganic chemical vapor deposition. *J. Appl. Phys.* **1999**, *85*, 3824–3831. [CrossRef]
23. Kubo, M.; Mannoh, M.; Narusawa, T. Raman study of crystalline structure and resonant behavior in $(Al_xGa_{1-x})_{0.5}In_{0.5}P$ quaternary alloys. *J. Appl. Phys.* **1989**, *66*, 3767–3775. [CrossRef]
24. Kondow, M.; Minagawa, S.; Satoh, S. Raman scattering from AlGaInP. *Appl. Phys. Lett.* **1987**, *51*, 2001–2003. [CrossRef]
25. Jia, P.; Qin, L.; Zhang, X.; Zhang, J.; Liu, T.-Y.; Men, Z.-W.; Ning, Y.-Q. Reliability Study of Grating Coupled Semiconductor Laser Based on Raman Spectra Technique. *Spectrosc. Spect. Anal.* **2016**, *36*, 1745–1748.
26. Spencer, B.J.; Voorhees, P.W.; Davis, S.H. Morphological instability in epitaxially strained dislocation-free solid films: Linear stability theory. *J. Appl. Phys.* **1993**, *73*, 4955–4970. [CrossRef]
27. Snyder, C.W.; Orr, B.G.; Kessler, D.; Sander, L.M. Effect of strain on surface morphology in highly strained InGaAs films. *Phys. Rev. Lett.* **1991**, *66*, 3032–3035. [CrossRef]
28. Watanabe, M.; Okajima, M. Dependence of InGaP/InGaAlP heterointerface smoothness on substrate misorientation. *J. Appl. Phys.* **1996**, *80*, 3291–3296. [CrossRef]
29. Brown, A.S.; Mishra, U.K.; Henige, J.A.; Delaney, M.J. The effect of InP substrate misorientation on GaInAs-AlInAs interface and alloy quality. *J. Appl. Phys.* **1988**, *64*, 3476–33480. [CrossRef]
30. Varshni, Y.P. Temperature dependence of the energy gap in semiconductors. *Physica* **1967**, *34*, 149–154. [CrossRef]
31. Chen, X.; Zhao, B.-J.; Ren, Z.-W.; Tong, J.-H.; Wang, X.-F.; Zhuo, X.-J.; Zhang, J.; Li, D.-W.; Yi, H.-X.; Li, S.-T.; et al. Improvement of light extraction efficiency in InGaN/GaN-based light-emitting diodes with a nano-roughened p-GaN surface. *J. Mater. Sci. Mater. Electron.* **2014**, *25*, 4200–4205.
32. De, S.; Layek, A.; Raja, A.; Kadir, A.; Gokhale, M.R.; Bhattacharya, A.; Dhar, S.; Chowdhury, A. Two distinct origins of highly localized luminescent centers within InGaN/GaN quantum-well light-emitting diodes. *Adv. Funct. Mater.* **2011**, *21*, 3828–3835. [CrossRef]
33. Wang, Y.; Pei, X.J.; Xing, Z.G.; Guo, L.W.; Jia, H.Q.; Chen, H.; Zhou, J.M. Effects of barrier growth temperature ramp-up time on the photoluminescence of InGaN/GaN quantum wells. *J. Appl. Phys.* **2007**, *101*, 033509. [CrossRef]
34. Olaizola, S.M.; Pendlebury, S.T.; O'Neill, J.P.; Mowbray, D.J.; Cullis, A.G.; Skolnick, M.S.; Parbrook, P.J.; Fox, A.M. Influence of GaN barrier growth temperature on the photoluminescence of InGaN/GaN heterostructures. *J. Phys. D* **2002**, *35*, 599–603. [CrossRef]
35. Bai, J.; Wang, T.; Sakai, S. Influence of the quantum-well thickness on the radiative recombination of InGaN/GaN quantum well structures. *J. Appl. Phys.* **2000**, *88*, 4729–4733. [CrossRef]
36. Eliseev, P.G. The red σ 2/kT spectral shift in partially disordered semiconductors. *J. Appl. Phys.* **2003**, *93*, 5404–5415. [CrossRef]
37. Wang, H.; Ji, Z.; Qu, S.; Wang, G.; Jiang, Y.; Liu, B.; Xu, X.; Mino, H. Influence of excitation power and temperature on photoluminescence in InGaN/GaN multiple quantum wells. *Opt. Express* **2012**, *20*, 3932–3940. [CrossRef] [PubMed]
38. Cho, Y.H.; Gainer, G.H.; Fischer, A.J.; Song, J.J.; Keller, S.; Mishra, U.K.; DenBaars, S.P. "S-shaped" temperature-dependent emission shift and carrier dynamics in InGaN/GaN multiple quantum wells. *Appl. Phys. Lett.* **1998**, *73*, 1370–1372. [CrossRef]
39. Sarkar, N.; Ghosh, S. The temperature dependence of the band gap shrinkage due to the electron–phonon interaction in $Al_xGa_{1-x}As$. *J. Phys. Condens. Matter* **2006**, *18*, 1687–1694. [CrossRef]
40. Zhu, T.; Ertekin, E. Resolving anomalous strain effects on two-dimensional phonon flows: The cases of graphene, boron nitride, and planar superlattices. *Phys. Rev. B* **2015**, *91*, 205429. [CrossRef]
41. Vashaee, D.; Shakouri, A. Electronic and thermoelectric transport in semiconductor and metallic superlattices. *J. Appl. Phys.* **2004**, *95*, 1233–1245. [CrossRef]
42. Rudin, S.; Reinecke, T.L. Temperature-dependent exciton linewidths in semiconductor quantum wells. *Phys. Rev. B* **1990**, *41*, 3017–3027. [CrossRef] [PubMed]
43. Song, Y.; Zhang, L.; Zeng, Y.; Chen, Y.; Qin, L.; Zhou, Y.; Luo, Y.; Zhao, H.; Ning, Y.; Wang, L. Temperature-dependent photoluminescence characterization of compressively strained AlGaInAs quantum wells. *Mater. Res. Bull.* **2019**, *115*, 196–200. [CrossRef]
44. Deng, Z.; Jiang, Y.; Ma, Z.; Wang, W.; Jia, H.; Zhou, J.; Chen, H. A novel wavelength-adjusting method in InGaN-based light-emitting diodes. *Sci. Rep.* **2013**, *3*, 3389. [CrossRef]
45. Fang, Y.; Wang, L.; Sun, Q.; Lu, T.; Deng, Z.; Ma, Z.; Jiang, Y.; Jia, H.; Wang, W.; Zhou, J.; et al. Investigation of temperature-dependent photoluminescence in multi-quantum wells. *Sci. Rep.* **2015**, *5*, 12718. [CrossRef] [PubMed]
46. Lu, T.; Ma, Z.; Du, C.; Fang, Y.; Wu, H.; Jiang, Y.; Wang, L.; Dai, L.; Jia, H.; Liu, W.; et al. Temperature-dependent photoluminescence in light-emitting diodes. *Sci. Rep.* **2014**, *4*, 6131. [CrossRef]
47. Liu, L.; Wang, L.; Li, D.; Liu, N.; Li, L.; Cao, W.; Yang, W.; Wan, C.; Chen, W.; Du, W.; et al. Influence of indium composition in the prestrained InGaN interlayer on the strain relaxation of InGaN/GaN multiple quantum wells in laser diode structures. *J. Appl. Phys.* **2011**, *109*, 073106. [CrossRef]
48. Ding, Y.J.; Guo, C.L.; Li, S.; Khurgin, J.B.; Law, K.K.; Merz, J.L. Continuous-wave photoluminescence excitation spectra of multiple narrow-stepped quantum wells: Evidence for saturation of interface traps. *Appl. Phys. Lett.* **1992**, *60*, 154–156. [CrossRef]

49. Hillmer, H.; Forchel, A.; Sauer, R.; Tu, C.W. Interface-roughness-controlled exciton mobilities in GaAs/$Al_{0.37}Ga_{0.63}$ As quantum wells. *Phys. Rev. B* **1990**, *42*, 3220–3223. [CrossRef]
50. Weber, S.; Limmer, W.; Thonke, K.; Sauer, R.; Panzlaff, K.; Bacher, G.; Meier, H.P.; Roentgen, P. Thermal carrier emission from a semiconductor quantum well. *Phys. Rev. B* **1995**, *52*, 14739–14747. [CrossRef]

applied sciences

MDPI

Article

100 Gbaud On–Off Keying/Pulse Amplitude Modulation Links in C-Band for Short-Reach Optical Interconnects

Oskars Ozolins [1,2,3,*], Xiaodan Pang [2], Aleksejs Udalcovs [1], Richard Schatz [2], Sandis Spolitis [3,4], Vjaceslavs Bobrovs [3], Gunnar Jacobsen [1] and Sergei Popov [2]

1 Networks Unit, RISE Research Institutes of Sweden, 164 40 Kista, Sweden; aleksejs.udalcovs@ri.se (A.U.); gjacobsen1952@gmail.com (G.J.)
2 Applied Physics Department, KTH Royal Institute of Technology, 106 91 Stockholm, Sweden; xiaodan@kth.se (X.P.); rschatz@kth.se (R.S.); sergeip@kth.se (S.P.)
3 Institute of Telecommunications, Riga Technical University, 1048 Riga, Latvia; sandis.spolitis@rtu.lv (S.S.); vjaceslavs.bobrovs@rtu.lv (V.B.)
4 Communication Technologies Research Center, Riga Technical University, 1048 Riga, Latvia
* Correspondence: oskars.ozolins@ri.se or ozolins@kth.se or oskars.ozolins@rtu.lv; Tel.: +37-120-011-193

Abstract: We experimentally evaluate the high-speed on–off keying (OOK) and four-level pulse amplitude modulation (PAM4) transmitter's performance in C-band for short-reach optical interconnects. We demonstrate up to 100 Gbaud OOK and PAM4 transmission over a 400 m standard single-mode fiber with a monolithically integrated externally modulated laser (EML) having 100 GHz 3 dB bandwidth with 2 dB ripple. We evaluate its capabilities to enable 800 GbE client-side links based on eight, and even four, optical lanes for optical interconnect applications. We study the equalizer's complexity when increasing the baud rate of PAM4 signals. Furthermore, we extend our work with numerical simulations showing the required received optical power (ROP) for a certain bit error rate (BER) for the different combinations of the effective number of bits (ENOB) and extinction ratio (ER) at the transmitter. We also show a possibility to achieve around 1 km dispersion uncompensated transmission with a simple decision feedback equalizer (DFE) for a 100 Gbaud OOK, PAM4, and eight-level PAM (PAM8) link having the received power penalty of around 1 dB.

Keywords: optical interconnects; on–off keying; pulse amplitude modulation

Citation: Ozolins, O.; Pang, X.; Udalcovs, A.; Schatz, R.; Spolitis, S.; Bobrovs, V.; Jacobsen, G.; Popov, S. 100 Gbaud On–Off Keying/Pulse Amplitude Modulation Links in C-Band for Short-Reach Optical Interconnects. *Appl. Sci.* **2021**, *11*, 4284. https://doi.org/10.3390/app11094284

Academic Editor: Giuseppe Rizzelli

Received: 19 March 2021
Accepted: 5 May 2021
Published: 9 May 2021

1. Introduction

Datacenters experience enormous traffic growth due to the vast amount of data to be stored, transmitted, and processed [1–5]. This explosive growth of Internet Protocol (IP) traffic is driving datacenters to the so-called "Zettabyte Era". The Cisco Report predicts that annual global IP traffic will reach over 4.8 zettabytes/year by 2022 [5]. For some applications, such as those on Facebook, the internal traffic may be several orders of magnitude greater than external. In addition to the internal traffic required to build web pages and search indices, relatively recent machine learning (ML) applications are driving increasing amounts of both computation and traffic on the datacenter interconnection network [4]. Due to that, challenges arise to keep up the bandwidth scalability. The community is looking into cost-efficient short-reach optical interconnect for 800 GbE intra-datacenter links [6]. We see solutions based on eight optical lanes thanks to compatibility with 400 GbE building blocks [7]. Eight optical lanes-based 400 GbE solutions are already being deployed [8,9]. Solutions based on eight optical lanes [10–15], or even four optical lanes [16–27], for 800 GbE are more appealing thanks to the use of using high bandwidth components. This allows reducing of costs, power consumption, and complexity of parallelism.

The current industrial solution for 400 GbE is based on four-level pulse amplitude modulation (PAM4) [8,9] instead of on–off keying (OOK) [19,28]. PAM4 reduces bandwidth requirements without excessive costs for sensitivity, digital signal processing (DSP), and transceiver complexity [29–32]. The higher spectral efficiency of advanced multilevel modulation formats, such as carrier-less amplitude and phase (CAP) or discrete multi-tone (DMT) modulations [33–36], can be beneficial to increase the bitrate of the single lane. However, we need to take into consideration the implementational complexity. Advanced modulation formats are also being extensively investigated to scale up the speed, even up to 1.6 TbE [37,38]. To gradually achieve this goal, we need to increase the baud rate of individual lanes even higher. In that manner, we can achieve the four optical lanes 800 GbE solution. That requires broadband InP and/or silicon optoelectronic components with high phase and amplitude linearity [39–42]. Optical interconnects operating at 100 Gbaud and beyond per lane are needed. In this way, we can confront the growing bandwidth density and scalability issues in datacenters. We foresee that the industrial solution for codesigned and cointegrated electronics and photonics is around the corner. The required ultrahigh bandwidth building blocks are available: integrated circuits for generating PAM4 [43,44] even without equalization [45]; modulators and photodetectors (PD) [17,39,46]; and analogue to digital converter (ADC) frontend for receivers [47]. We also need to mention that electronic instruments, such as arbitrary waveform generators (AWG) and digital storage oscilloscopes (DSO), have reached sampling rates beyond 200 GSa/s with plenty of bandwidth for 100 Gbaud signals. This will allow scaling bandwidth for intra-datacenter networks in a cost-efficient manner. The C-band enables proportional efficiency in bidirectional bandwidth for intra-datacenter links thanks to mature and efficient wavelength division multiplexing technology. However, the reach is limited in this band due to the chromatic dispersion (CD) imposing limitations for the system design. It is worth looking into the achievable system performance and reach with high baud rate to enable four optical lanes solutions for 800 GbE links.

In this paper, we experimentally evaluate high-speed OOK and PAM4 transmitter performance in C-band for short-reach optical interconnect applications. Compared to our previous demonstrations [18,20], we show detailed analysis on equalizer configurations and filtering for signals operating at up to 100 Gbaud. It is important to show the pros and the cons of the PAM4 generation scheme used to achieve this result. We characterize the performance using the externally modulated laser (EML) having 100 GHz 3 dB bandwidth with a 2 dB ripple. Furthermore, we also include the results of numerical simulations. We study required received optical power (ROP) for a certain bit error rate (BER) for the different effective number of bits (ENOB) and extinction ratio (ER) at the transmitter. This allows estimating what implementation penalty to expect when designing high-speed transmitters. Besides, we also show achievable transmission reach in C-band for 100 Gbaud OOK, PAM4, and eight-level PAM (PAM8) signals over standard single-mode fiber with a simple decision feedback equalizer (DFE). This paves the way for cost-effective short-reach optical interconnects with high-speed multilevel modulation formats. The remaining parts of the paper are organized as follows. In Section 2, the experiment setup is detailed, and corresponding results and discussion are presented. In Section 3, we show the simulation setup and results along with a discussion. The conclusions are drawn in Section 4.

2. Experimental Setup and Results

Figure 1a shows the experimental setup for evaluating the intra-datacenter link performance. First, we amplify and decorrelate two pseudorandom bit sequences with a word length of 2^{15}-1 (PRBS15) at 40 Gbaud, 45 Gbaud, and 50 Gbaud. Then, we multiplex them in time domain using 2:1 selector [48] to form a single 80 Gbaud, 90 Gbaud, and 100 Gbaud nonreturn to zero (NRZ) sequences, respectively. In the case of OOK, we use one of the multiplexed outputs (as shown in Figure 1 with dashed line) at 100 Gbaud amplified in a 65 GHz linear amplifier to drive the EML. In the case of PAM4, the attenuated signal is passively combined in an electrical 65 GHz three-resistor combiner with an inverted decorrelated version. The generated PAM4 signal at electrical back-to-back (b2b) already has a penalty due to the nonequally spaced signal amplitude levels. This happens due to different losses in two paths of OOK signals that do not allow to achieve a precise 6 dB difference, since the decorrelation suffers from technical implementation limitations.

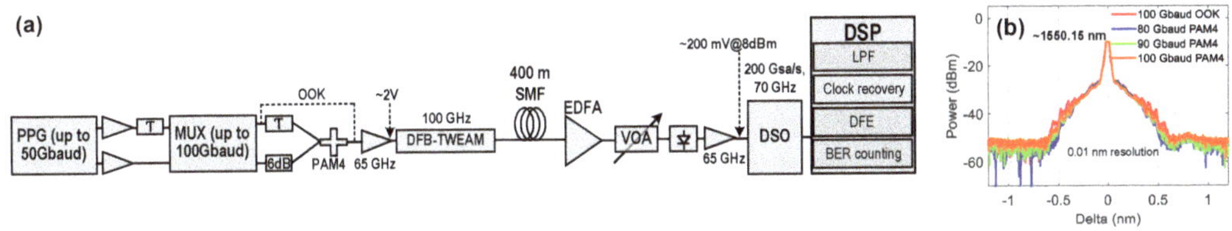

Figure 1. (**a**) Experimental setup (PPG: pulse-pattern generator), (**b**) optical spectra of the 100 Gbaud OOK and PAM4 signals (Mux: multiplexer, T: delay line, DFB-TWEAM: distributed feedback laser with traveling-wave electroabsorption modulator, SMF: single-mode fiber, EDFA: erbium-doped fiber amplifier, VOA: variable optical attenuator, DSO: digital storage oscilloscope, DSP: digital signal processing, LPF: low-pass filter, DFE: decision feedback equalizer, BER: bit error rate).

In this paper, we wish to quantify this simple PAM4 generation scheme's impact on the introduced penalty. Therefore, we gradually increase baud rate from 80 Gbaud to 100 Gbaud. Then, the PAM4 signal is amplified in a 65 GHz linear amplifier to drive the EML. We achieve around 2 volts of the electrical signal (peak to peak) for OOK and PAM4 high-speed signals. EML is based on a monolithically integrated distributed feedback laser with traveling-wave electroabsorption modulator (DFB-TWEAM) designed by KTH, fabricated by KTH and Syntune, and packaged by u^{2}t Photonics. The DFB laser threshold current is around 25 mA and the slope efficiency is 0.04 W/A. The wavelength of the DFB-TWEAM is 1550.15 nm in this experiment. The EML has a 3 dB bandwidth beyond 100 GHz with less than 2 dB ripple in the passband, which indicates high phase linearity [39]. These figures of merits make this EML considered an excellent candidate for four optical lanes-based 800 GbE solution [19,20,36]. We use a current of 120 mA for the DFB and a voltage of −1.85 V for the TWEAM, resulting in around a 5 dB extinction ratio. We achieve an output power of −0.9 dBm. Higher power values can be achieved with a reduced bias voltage, which also reduces the extinction ratio of a modulated signal. In the experiment, we found a trade-off between the bias voltage and the extinction ratio. In Figure 1b, one can see that the modulated optical spectra are around 1550.15 nm for the OOK and PAM4. The optical link consists of 400 m of SMF. The receiver consists of a preamplifier, a variable optical attenuator (VOA), a >100 GHz PD from u^{2}t with a responsivity of 0.5 A/W, a 65 GHz linear amplifier, and a 200 GSa/s, a 70 GHz bandwidth Tektronix DSO (DPO77002SX). An automatic gain-controlled erbium-doped fiber amplifier (EDFA) with fixed output power is employed as the preamplifier due to a low PD responsivity. The noise figure of the amplifier is around 4.5 dB. Due to limited maximum input power to the photodetector, an additional 65 GHz linear amplifier is used to ensure around 200 mV peak to peak voltage of the received signals required for better performance. In the case of OOK, we could skip the optical amplification when no variable optical attenuator is used. Instead, we keep the amplifier for OOK to compare it to PAM4 under similar noise conditions. The sampled signal is then processed offline using a typical DSP. We perform low-pass filtering, clock recovery, and resampling to ensure 1 sample per symbol for a symbol-spaced DFE with a different configuration of feed-forward taps (FFT) and feedback taps (FBT) to overcome inter symbol interference (ISI) in the presence of the noise. The main limitations are due to the limited effective 3 dB bandwidth of the electrical components and the implementation penalty itself. After the equalization, we perform BER counting. We consider hard-decision forward error correction code (HD-FEC) with 7% and 20% overheads (OH) and soft-decision FEC (SD-FEC) with 20% OH with pre-FEC BERs at 5E-3 [49], 1.1E-2 [50], and 2E-2, respectively. We use the SD-FEC due to the poor electrical b2b signal quality related to the implementation penalty. We study different DFE configurations to improve the received signal quality (see Figure 2). We express the performance in terms of ROP required for a certain BER. We use a 7% HD-FEC limit of 5E-3 and a 20% SD-FEC limit of 2E-2 for OOK and PAM4 signals, respectively. For both signals, we fix the number of feedback taps to 15 and a low-pass filter (LPF) bandwidth to 0.75*baud rate. Then we choose the feed-forward taps number to be 3, 7, 11, and 15 for OOK and 15, 29, 43, 57, and 71 for PAM4 signals.

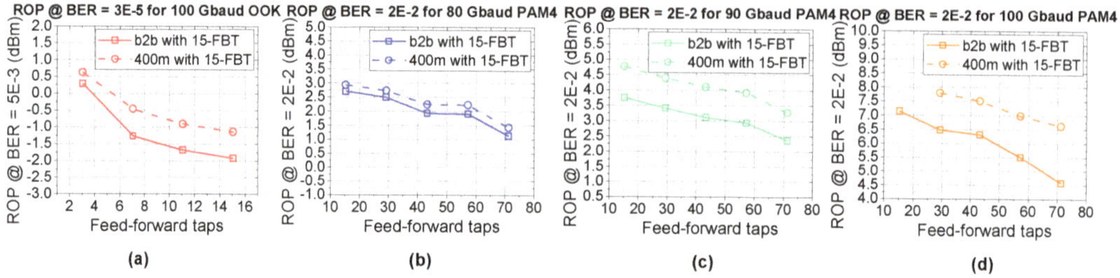

Figure 2. Received optical power versus different configurations of decision feedback equalizers for (**a**) 100 Gbaud OOK, (**b**) 80 Gbaud PAM4, (**c**) 90 Gbaud PAM4, and (**d**) 100 Gbaud PAM4 signals.

From Figure 2, we observe that with the increase in the feed-forward tap number we require lower received optical power to achieve the specified pre-FEC BER. For 100 Gbaud OOK signal with 15-tap FFT and 15-tap FBT equalizer, we see around 0.7 dB power penalty for 400 m SMF transmission compared to optical back-to-back. If we compare PAM4 signals with the same equalizer, we get a power penalty of around 0.2 dB for 80 Gbaud and 1 dB for 90 Gbaud signals. For 100 Gbaud PAM4 signal after 400 m transmission over the SMF, we could not achieve a bit error rate of 2E-2 for any received power. This forced us to increase the complexity of the equalizer. Still, we observe a higher power penalty—around 1.3 dB with 29-tap FFT and 15-tap FBT equalizer. For further analysis, we choose equalizers with higher complexity to reduce the required ROP for a certain BER.

Afterward, we obtain the bit error rate as a function of LPF_{BW}/baud rate for OOK and PAM4 signals (see Figure 3). During the processing, we set the LPF_{BW}/baud rate from 0.45 to 0.8. We choose received optical power of −1 dBm for OOK and 8 dBm for PAM4 signals to observe the BER performance around the specific FEC limits mentioned above. For the OOK signal, both curves are mostly below the 7% HD-FEC limit. Similar performance for optical b2b and after 400 m is obtained with a 15-tap FFT and 15-tap FBT equalizer. We see that the best signal performance is for the LPF_{BW}/baud rate of 0.575, which is now chosen for further processing. For PAM4 signals, we must increase the equalizer's complexity. We increase the number of feed-forward taps to 71 and keep the same number of feedback taps, i.e., FBT = 15. For 80 Gbaud PAM4 signal, we managed to reach BERs below 5E-3 for above 0.6 of LPF_{BW}/baud rate for both curves. It is also the case for the optical b2b curve of 90 Gbaud PAM4 signals with a 71-tap FFT and 15-tap FBT equalizer and LPF_{BW}/baud rate of 0.75. We need to increase the HD-FEC overhead to 20% to detect 90 Gbaud PAM4 signals after 400 m of SMF transmission. In the case of 100 Gbaud PAM4, the performance is below the 20% SD-FEC limit. For all three cases of the PAM4 signals, we choose LPF_{BW}/baud rate of 0.75 for further analysis.

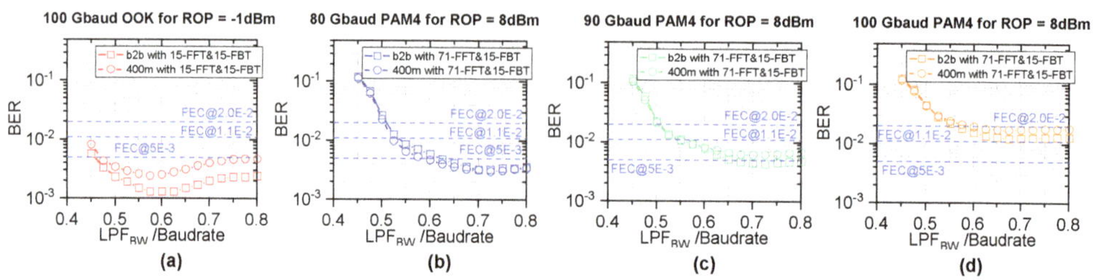

Figure 3. Bit error rate versus LPF_{bw}/baud rate for (**a**) 100 Gbaud OOK, (**b**) 80 Gbaud PAM4, (**c**) 90 Gbaud PAM4, and (**d**) 100 Gbaud PAM4 signals. We use minus 1 dBm received optical power for OOK and 8 dBm for PAM4 signals.

Finally, we show bit error rate, eye-diagrams, and amplitude histograms for OOK and PAM4 signals after the optical b2b and 400 m transmission. BER results as a function of received power for different signals and DFE configurations are shown in Figure 4a. The BER curves are obtained using a 15-FFT and 15-FBT equalizer for OOK and a 71-FFT and 15-FBT for PAM4. In the case of OOK, we manage to achieve the 7% HD-FEC limit for both the optical b2b and the 400-m transmission. We observe around 0.5 dB received power penalty at the 7% HD-FEC limit due to a large implementation penalty which adds up to the transmission penalty over 400 m of SMF. The implementation penalty can also be observed from the equalized eye diagram and histograms in Figure 4b. Further, we study transmission with 80 Gbaud PAM4 signals. One can observe a severe PAM4 signal degradation due to both higher sensitivity requirements and poor electrical signal performance at the transmitter due to the implementation penalty. This can be seen from the eye diagrams and amplitude histograms in Figure 4. The suboptimal passive combining ratio due to the imperfect components explains such worse performance. The electrical signal performance imposes a strict error floor for PAM4 signals. One can observe that the use of a 71-FFT and 15-FBT equalizer allows achieving below the 7% HD-FEC performance limit after 400 m of SMF for 80 Gbaud PAM4 signals with a small penalty compared to the optical b2b. We need to increase the number of equalizer feed-forward taps significantly compared to the OOK case to reduce the impact of the implementation penalty on high-speed PAM4 signals. When increasing the PAM4 signal data rate to 90 Gbaud, we obtain the performance below the HD-FEC limit only for optical b2b. The performance has degraded after 400 m transmission and we need to increase the overhead to 20% for the HD-FEC. In this case, we observe around 1.3 dB received power penalty. Then, we increase the transmission speed to 100 Gbaud and achieve only the 20% SD-FEC limit. Here, we observe around 2.3 dB of received power penalty. We achieve 149 Gbit/s, 150 Gbit/s, and 166 Gbit/s post-FEC bitrates for 80 Gbaud, 90 Gbaud, and 100 Gbaud PAM4 signals and 93 Gbit/s for 100 Gbaud OOK signal.

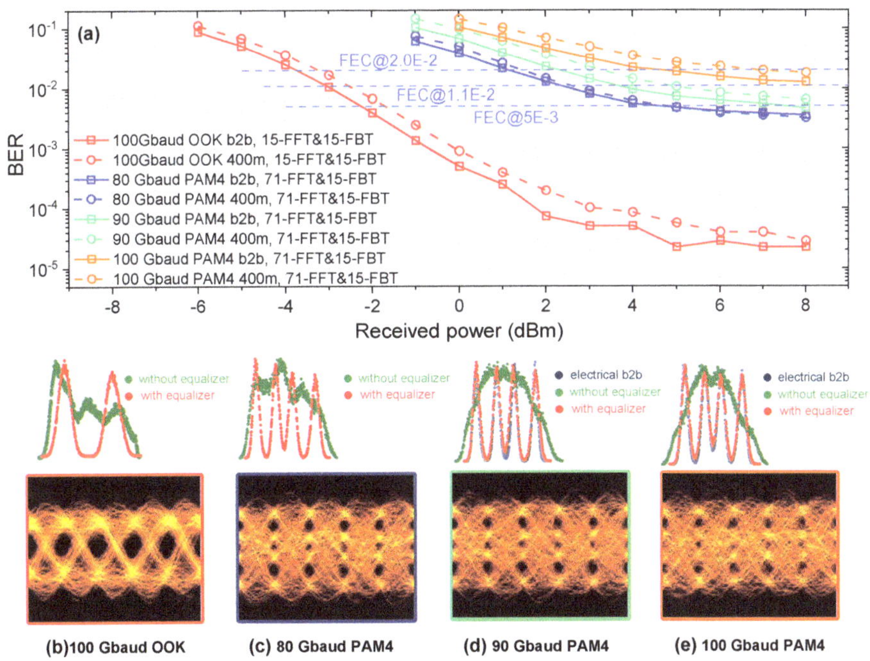

Figure 4. (**a**) BER curves for OOK and PAM4 signal for b2b and 400 m, (**b–e**) eye diagram and histograms for optical b2b at ROP of 8 dBm.

We point out that the histograms obtained with the 71-FFT and 15-FBT DFE for the electrical b2b and the optical b2b signals at 90 Gbaud and 100 Gbaud are comparable in their performance (see Figure 4d,e). We measure the electrical b2b signals before the DFB-TWEAM and after the 65 GHz amplifier. We attribute the main implementation penalty to electrical components in the transmitter which could be remedied by improving the bandwidth and linearity in the electrical domain. Further improvement in the electrical signal generation scheme for PAM potentially would allow achieving better signal quality and, thus, improving the link latency thanks to a reduced FEC [21,31]. This would result in an even higher baud rate for multilevel signal transmission with the EML [19]. The operational wavelength of the EML is around 1550 nm in the measurements. The obtained results demonstrate the capability of the EML to enable 800 GbE client-side links for short-reach optical interconnect applications.

3. Simulation Setup and Results

We perform numerical simulations to complement the experiments. We use the MATLAB-based Robochameleon framework [51]. It is a coding framework and component library for simulation and experimental analysis of optical communication systems. We focus on transmitter imperfections and CD tolerance. The simulation setup is shown in Figure 5 and its main configuration parameters are specified in Table 1. The transmitter includes a digital to analog converter (DAC), an intensity modulator (IM), and a continuous wave (CW) laser. The link is based on lossless single-mode fiber. In the receiver, we have a variable optical attenuator, a photodetector with a transimpedance amplifier (PD-TIA), and an ADC. Then we perform a typical DSP to recover transmitted bits. We choose a simple DFE with 6-FFT and 3-FBT to perform channel equalization since the modulator is driven in the linear region and we have a linear dispersive fiber link. We use hard-decision demodulators for the 100 Gbaud OOK, PAM4, and PAM8 signals after the equalization. We set a CD coefficient to 16 ps/nm/km. The PD has a bandwidth of 112 GHz and the Gaussian receiver filter has bandwidth of 75 GHz, as can be seen from Table 1. We use root-raised cosine pulse shaping with 0.75 roll-off to have similar signals as in the experiment.

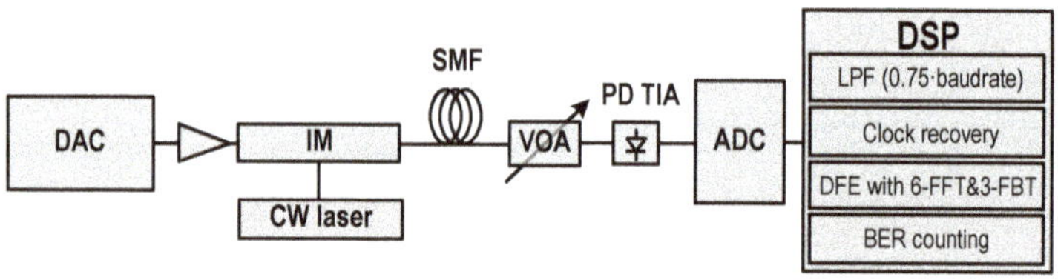

Figure 5. Simulation setup. DAC: digital to analog converter, CW laser: continuous-wave laser, IM: intensity modulator, SMF: single-mode fiber, VOA: variable optical attenuator, PD-TIA: photodetector with a transimpedance amplifier, ADC: analog to digital converter, DSP: digital signal processing, LPF: low-pass filter, DFE: decision feedback equalizer, FFT: feed-forward taps, FBT: feedback taps, BER: bit error rate.

We study the required ROP for a certain BER (see Figure 6a–c) and the power penalty (see Figure 6d–f) for the different ENOB and ER at the transmitter for 100 Gbaud OOK, PAM4, and PAM8 signals for optical b2b. The power penalty was calculated by using separate references for each modulation format. For the OOK and PAM4 signals, we sweep the ENOB from 2 to 6, while for the PAM8 from 3 to 7. This is because PAM8 has more amplitude levels to be represented. For all three modulation formats, we choose the extinction ratio to be from 3 to 10 dB. Values are color-coded with the respect to the minimum and the maximum ROP and penalty values, respectively. One can observe from the curves that the extinction ratio of the signal at the transmitter has a crucial impact on the required ROP for a certain BER and the power penalty. In all numerical simulations, we are using the 7% HD-FEC limit with a BER of 5E-3. One can see that the required ROP is modulation format-dependent. For instance, the required ROP is around minus 4.5 dBm for OOK, minus 1 dBm for PAM4, and 2 dBm for PAM8 when the ER is set to 5 dB and the ENOB to 5. We observe around 4 dB power penalty for OOK and PAM4 signals for the same ER and ENOB. The penalty is slightly higher for PAM8. For further simulations, we set ER = 5 dB and ENOB = 5, because it is the closest value to the experiment.

Table 1. Simulation setup parameters.

Transmitter	Parameters
Modulation format	OOK, PAM4, PAM8
Pulse shaping	Root raised cosine
Roll-off	0.75
ENOB	2 to 7 or 5
Modulator driving signal Vpp	0.4*Vπ
Laser RIN	145 dB/Hz
Extinction ratio	3 to 10 or 5
Output power	5 dBm
Wavelength	1550 nm
Lossless Channel	**Parameters**
Dispersion	16 ps/nm
Length	0 to 1000 m
Receiver	**Parameters**
Photodetector	PIN-TIA
Photodetector responsivity	0.5 A/W
Noise current	18 pA/$\sqrt{}$Hz
Photodetector filter model	4th order Butterworth filter
PD-TIA cut-off bandwidth	112 GHz
Digital Signal Processing	**Parameters**
Gaussian filter bandwidth	0.75*baud rate
Equalizer	Decision feedback equalizer
Demodulation	OOK/PAM
FEC limit	5E-3

V_{PP}: peak-to-peak voltage; V_{π} = 2 V.

Fiber's CD will limit the transmission distance if we consider single-mode operations in the C-band. Therefore, we also investigate CD tolerance for different modulation formats. We are not considering chirp at the transmitter to show degradation coming only from fiber's dispersion at the operational wavelength of 1550 nm. The received power penalty as a function of transmission distance is shown in Figure 7. We obtain curves for a 6-FFT and 3-FBT equalizer. Considering a 1 dB power penalty, the 100 Gbaud PAM4 can be transmitted up to 1000 m. It is more than two times (2×) increased compared to the experimental demonstration. For the 100 Gbaud OOK, the fiber distance can be increased beyond 1000 m while maintaining the same penalty and equalizer parameters. The transmission distances over SMF can be significantly improved by the microwave design of the EML that can be applied to a semiconductor material with a larger bandgap to achieve modulation at zero dispersion wavelength [39].

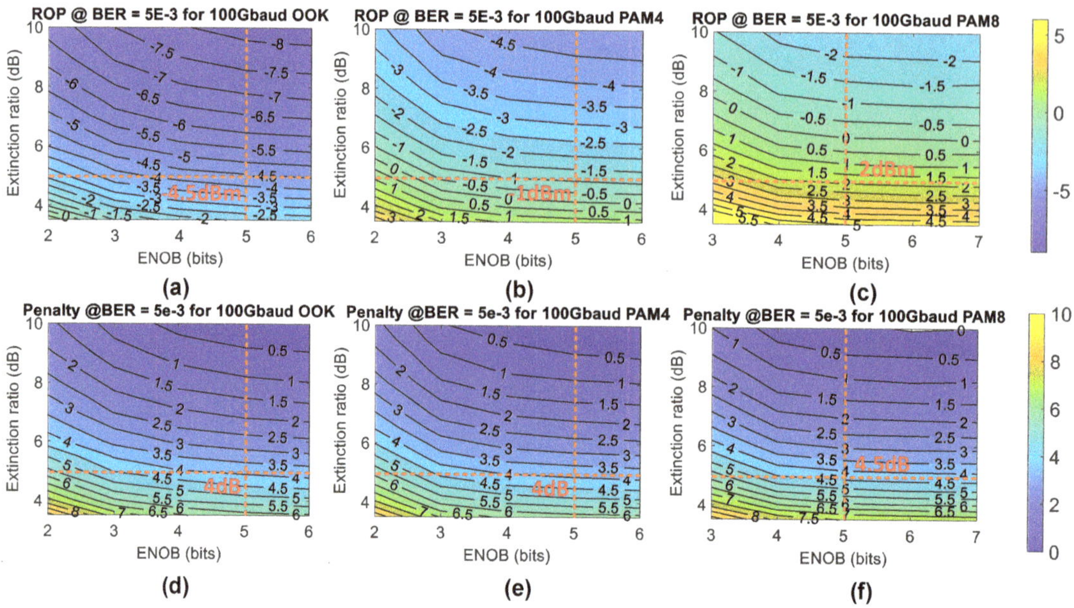

Figure 6. Extinction ratio versus the effective number of bits in terms of received optical power and penalty for 100 Gbaud (**a**,**d**) OOK, (**b**,**e**) 8PAM4, and (**c**,**f**) PAM8 signals for 6-FFT and 3FBT equalizer configuration at optical b2b.

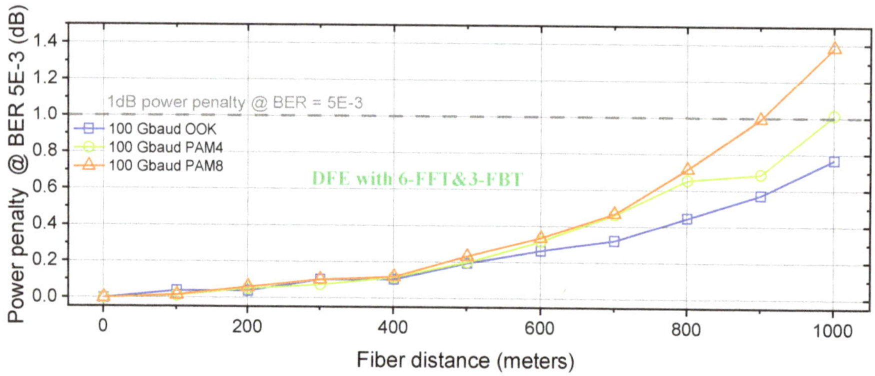

Figure 7. Received power penalty (at BER = 5E-3) as a function of transmitted distance for 100 Gbaud OOK, PAM4, and PAM8 signals with 6-FFT and 3FBT equalizer configuration (see inset).

4. Conclusions

We have demonstrated 100 Gbaud OOK and PAM4 transmitter performance in C-band for short-reach optical interconnect applications as a potential solution for four/eight optical lanes 800 GbE links. We achieve 149 Gbit/s, 150 Gbit/s, and 166 Gbit/s post-FEC bitrates for 80 Gbaud, 90 Gbaud, and 100 Gbaud PAM4 signals and 93 Gbit/s for 100 Gbaud OOK signal. Furthermore, we perform numerical simulations to study the impact of transmitter parameters and chromatic dispersion impact of power penalty for 100 Gbaud OOK, PAM4, and PAM8 signals. We observe the significant chromatic dispersion impact in both experiments and simulations.

Author Contributions: Conceptualization, O.O., X.P., A.U.; methodology, O.O., X.P., A.U., R.S., V.B., S.P.; software, O.O., X.P.; validation, O.O., X.P., A.U., S.S.; formal analysis, O.O., X.P., A.U., R.S., S.S.; investigation, O.O., X.P., A.U., R.S.; resources, O.O.; data curation, O.O., X.P., A.U., R.S., S.S.; writing—original draft preparation, O.O.; writing—review and editing, X.P., A.U., R.S., S.S., V.B., G.J., S.P.; visualization, O.O., X.P., A.U.; supervision, V.B., G.J., S.P.; project administration, O.O., R.S.; funding acquisition, O.O., R.S., V.B., S.P. All authors have read and agreed to the published version of the manuscript.

Funding: This work was supported by the ERDF-funded CARAT project No. 1.1.1.2/VIAA/4/20/660 and the EU RIA TWILIGHT project no. 871741.

Institutional Review Board Statement: Not applicable.

Informed Consent Statement: Not applicable.

Data Availability Statement: Not applicable.

Acknowledgments: We acknowledge Tektronix for lending us the 200 GSa/s 70 GHz oscilloscope (DPO77002SX).

Conflicts of Interest: The authors declare no conflict of interest.

References

1. Mestre, M.A.; Mardoyan, H.; Caillaud, C.; Rios-Müller, R.; Renaudier, J.; Jennevé, P.; Blache, F.; Pommereau, F.; Decobert, J.; Jorge, F.; et al. Compact InP-based DFB-EAM enabling PAM-4 112 Gb/s transmission over 2 km. *J. Lightwave Technol.* **2016**, *34*, 1572–1578. [CrossRef]
2. Mardoyan, H.; Mestre, M.A.; Rios-Müller, R.; Konczykowska, A.; Renaudier, J.; Jorge, F.; Duval, B.; Dupuy, J.; Ghazisaeidi, A.; Jennevé, P.; et al. Single carrier 168-Gb/s line-rate PAM direct detection transmission using high-speed selector power DAC for optical interconnects. *J. Lightwave Technol.* **2016**, *34*, 1593–1598. [CrossRef]
3. Xiang, M.; Xing, Z.; El-Fiky, E.; Morsy-Osman, M.; Zhuge, Q.; Plant, D.V. Single-lane 145 Gbit/s IM/DD transmission with faster-than-nyquist PAM4 signaling. *IEEE Photonics Technol. Lett.* **2018**, *30*, 1238–1241. [CrossRef]
4. Cheng, Q.; Bahadori, M.; Glick, M.; Rumley, S.; Bergman, K. Recent advances in optical technologies for data centers: A review. *Optica* **2018**, *5*, 1354–1370. [CrossRef]
5. Cisco Inc. *Cisco Annual Internet Report (2018–2023)*; Cisco: San Jose, CA, USA, 2020.
6. Sun, H.; Torbatian, M.; Karimi, M.; Maher, R.; Thomson, S.; Tehrani, M.; Gao, Y.; Kumpera, A.; Soliman, G.; Kakkar, A.; et al. 800G DSP ASIC design using probabilistic shaping and digital sub-carrier multiplexing. *J. Lightwave Technol.* **2020**, *38*, 4744–4756. [CrossRef]
7. Karinou, F.; Prodaniuc, C.; Stojanovic, N.; Ortsiefer, M.; Daly, A.; Hohenleitner, R.; Kögel, B.; Neumeyr, C. Directly PAM-4 modulated 1530-nm VCSEL enabling 56 Gb/s/λ data-center interconnects. *IEEE Photonics Technol. Lett.* **2015**, *27*, 1872–1875. [CrossRef]
8. Filer, M.; Searcy, S.; Fu, Y.; Nagarajan, R.; Tibuleac, S. Demonstration and performance analysis of 4 Tb/s DWDM metro-DCI system with 100G PAM4 QSFP28 modules. In *Optical Fiber Communication Conference, Proceedings of the Optical Fiber Communication Conference, Los Angeles, CA, USA, 19–23 March 2017*; Optical Society of America: Los Angeles, CA, USA, 2017; p. W4D.4.
9. Nagarajan, R.; Filer, M.; Fu, Y.; Kato, M.; Rope, T.; Stewart, J. Silicon photonics-based 100 Gbit/s, PAM4, DWDM data center interconnects. *IEEE/OSA J. Opt. Commun. Netw.* **2018**, *10*, 25–36. [CrossRef]
10. Matsui, Y.; Schatz, R.; Pham, T.; Ling, W.A.; Carey, G.; Daghighian, H.M.; Adams, D.; Sudo, T.; Roxlo, C. 55 GHz bandwidth distributed reflector laser. *J. Lightwave Technol.* **2017**, *35*, 397–403. [CrossRef]
11. Pang, X.; Ozolins, O.; Gaiarin, S.; Kakkar, A.; Navarro, J.R.; Olmedo, M.I.; Schatz, R.; Udalcovs, A.; Westergren, U.; Zibar, D.; et al. Experimental study of 1.55-μm EML-based optical IM/DD PAM-4/8 short reach systems. *IEEE Photonics Technol. Lett.* **2017**, *29*, 523–526. [CrossRef]
12. Zhong, K.; Zhou, X.; Wang, Y.; Wang, Y.; Zhou, W.; Chen, W.; Zeng, L.; Wang, L.; Lau, A.P.T.; Lu, C. Transmission of a 120-GBd PM-NRZ signal using a monolithic double-side EML. *IEEE Photonics Technol. Lett.* **2016**, *28*, 2176–2179. [CrossRef]
13. Samani, A.; Chagnon, M.; El-Fiky, E.; Patel, D.; Jacques, M.; Veerasubramanian, V.; Plant, D.V. Silicon photonics modulator architectures for multi-level signal generation and transmission. In Proceedings of the 2017 Optical Fiber Communications Conference and Exhibition (OFC), Los Angeles, CA, USA, 19–23 March 2017; pp. 1–3.
14. Verplaetse, M.; Lin, R.; Kerrebrouck, J.V.; Ozolins, O.; Keulenaer, T.D.; Pang, X.; Pierco, R.; Vaernewyck, R.; Vyncke, A.; Schatz, R.; et al. Real-time 100 Gb/s transmission using three-level electrical duobinary modulation for short-reach optical interconnects. *J. Lightwave Technol.* **2017**, *35*, 1313–1319. [CrossRef]
15. Zhong, K.; Mo, J.; Grzybowski, R.; Lau, A.P.T. 400 Gbps PAM-4 signal transmission using a monolithic laser integrated silicon photonics transmitter. In Proceedings of the 2019 Optical Fiber Communications Conference and Exhibition (OFC), San Diego, CA, USA, 3–7 March 2019; pp. 1–3.

16. Kanazawa, S.; Yamazaki, H.; Nakanishi, Y.; Ueda, Y.; Kobayashi, W.; Muramoto, Y.; Ishii, H.; Sanjoh, H. 214-Gb/s 4-PAM operation of flip-chip interconnection EADFB laser module. *J. Lightwave Technol.* **2017**, *35*, 418–422. [CrossRef]
17. Lange, S.; Wolf, S.; Lutz, J.; Altenhain, L.; Schmid, R.; Kaiser, R.; Schell, M.; Koos, C.; Randel, S. 100 GBd intensity modulation and direct detection with an InP-based monolithic DFB laser Mach–Zehnder modulator. *J. Lightwave Technol.* **2018**, *36*, 97–102. [CrossRef]
18. Ozolins, O.; Pang, X.; Udalcovs, A.; Schatz, R.; Westergren, U.; Navarro, J.R.; Kakkar, A.; Nordwall, F.; Engenhardt, K.M.; Chen, J.; et al. 100 Gbaud 4PAM link for high speed optical interconnects. In Proceedings of the 2017 European Conference on Optical Communication (ECOC), Gothenburg, Sweden, 17–21 September 2017; pp. 1–3.
19. Estarán, J.M.; Mardoyan, H.; Jorge, F.; Ozolins, O.; Udalcovs, A.; Konczykowska, A.; Riet, M.; Duval, B.; Nodjiadjim, V.; Dupuy, J.; et al. 140/180/204-Gbaud OOK transceiver for inter- and intra-data center connectivity. *J. Lightwave Technol.* **2019**, *37*, 178–187. [CrossRef]
20. Pang, X.; Ozolins, O.; Lin, R.; Zhang, L.; Udalcovs, A.; Xue, L.; Schatz, R.; Westergren, U.; Xiao, S.; Hu, W.; et al. 200 Gbps/Lane IM/DD technologies for short reach optical interconnects. *J. Lightwave Technol.* **2020**, *38*, 492–503. [CrossRef]
21. Ozolins, O.; Zhang, L.; Udalcovs, A.; Louchet, H.; Dippon, T.; Gruen, M.; Pang, X.; Schatz, R.; Westergren, U.; Xiao, S.; et al. 100 Gbaud PAM4 link without EDFA and post-equalization for optical interconnects. In Proceedings of the 45th European Conference on Optical Communication (ECOC 2019), Dublin, Ireland, 22–26 September 2019; pp. 1–4.
22. Mardoyan, H.; Mestre, M.A.; Estarán, J.M.; Jorge, F.; Blache, F.; Angelini, P.; Konczykowska, A.; Riet, M.; Nodjiadjim, V.; Dupuy, J.; et al. 84-, 100-, and 107-GBd PAM-4 intensity-modulation direct-detection transceiver for datacenter interconnects. *J. Lightwave Technol.* **2017**, *35*, 1253–1259. [CrossRef]
23. Yang, C.; Li, W.; Yu, S. Single channel 224 Gbit/s (56-GBaud) PAM-16 transmission using linear digital pre-distortion. *Electron. Lett.* **2017**, *53*, 1420–1422. [CrossRef]
24. Chagnon, M.; Lessard, S.; Plant, D.V. 336 Gb/s in direct detection below KP4 FEC threshold for intra data center applications. *IEEE Photonics Technol. Lett.* **2016**, *28*, 2233–2236. [CrossRef]
25. Wei, J.; Rahman, T.; Calabrò, S.; Stojanovic, N.; Zhang, L.; Xie, C.; Ye, Z.; Kuschnerov, M. Experimental demonstration of advanced modulation formats for data center networks on 200 Gb/s lane rate IMDD links. *Opt. Express* **2020**, *28*, 35240–35250. [CrossRef] [PubMed]
26. Estarán, J.M.; Almonacil, S.; Rios-Müller, R.; Mardoyan, H.; Jenneve, P.; Benyahya, K.; Simonneau, C.; Bigo, S.; Renaudier, J.; Charlet, G. Sub-baudrate sampling at DAC and ADC: Toward 200G per lane IM/DD systems. *J. Lightwave Technol.* **2019**, *37*, 1536–1542. [CrossRef]
27. Che, D.; Matsui, Y.; Schatz, R.; Rodes, R.; Khan, F.; Kwakernaak, M.; Sudo, T. 200-Gb/s direct modulation of a 50-GHz class laser with advanced digital modulations. *J. Lightwave Technol.* **2021**, *39*, 845–852. [CrossRef]
28. Ozolins, O.; Estaran, J.M.; Udalcovs, A.; Jorge, F.; Mardoyan, H.; Konczykowska, A.; Riet, M.; Duval, B.; Nodjiadjim, V.; Dupuy, J.; et al. 140 Gbaud on-off keying links in C-band for short-reach optical interconnects. In Proceedings of the 2018 European Conference on Optical Communication (ECOC), Rome, Italy, 23–27 September 2018; pp. 1–3.
29. Zhang, Q.; Stojanovic, N.; Prodaniuc, C.; Xie, C.; Koenigsmann, M.; Laskowski, P. Single-lane 180 Gbit/s PAM-4 signal transmission over 2 km SSMF for short-reach applications. *Opt. Lett.* **2016**, *41*, 4449–4452. [CrossRef] [PubMed]
30. Zhong, K.; Zhou, X.; Huo, J.; Yu, C.; Lu, C.; Lau, A.P.T. Digital signal processing for short-reach optical communications: A review of current technologies and future trends. *J. Lightwave Technol.* **2018**, *36*, 377–400. [CrossRef]
31. Ozolins, O.; Pang, X.; Udalcovs, A.; Lin, R.; Kerrebrouck, J.V.; Gan, L.; Zhang, L.; Tang, M.; Fu, S.; Schatz, R.; et al. 7 × 149 Gbit/s PAM4 transmission over 1 km multicore fiber for short-reach optical interconnects. In Proceedings of the 2018 Conference on Lasers and Electro-Optics (CLEO), San Jose, CA, USA, 13–18 May 2018; pp. 1–2.
32. Chagnon, M. Optical communications for short reach. *J. Lightwave Technol.* **2019**, *37*, 1779–1797. [CrossRef]
33. Olmedo, M.I.; Zuo, T.; Jensen, J.B.; Zhong, Q.; Xu, X.; Popov, S.; Monroy, I.T. Multiband carrierless amplitude phase modulation for high capacity optical data links. *J. Lightwave Technol.* **2014**, *32*, 798–804. [CrossRef]
34. Zhang, L.; Hong, X.; Pang, X.; Ozolins, O.; Udalcovs, A.; Schatz, R.; Guo, C.; Zhang, J.; Nordwall, F.; Engenhardt, K.M.; et al. Nonlinearity-aware 200Gbit/s DMT transmission for C-band short-reach optical interconnects with a single packaged electro-absorption modulated laser. *Opt. Lett.* **2018**, *43*, 182–185. [CrossRef] [PubMed]
35. Ozolins, O.; Zhang, L.; Udalcovs, A.; Louchet, H.; Dippon, T.; Gruen, M.; Pang, X.; Schatz, R.; Westergren, U.; Xiao, S.; et al. 300+ Gbps short-reach optical communications. In Proceedings of the 2020 Conference on Lasers and Electro-Optics (CLEO), Washington, DC, USA, 10–15 May 2020; pp. 1–2.
36. Zhang, L.; Chen, J.; Udalcovs, A.; Pang, X.; Schatz, R.; Westergren, U.; Popov, S.; Xiao, S.; Ozolins, O. Kernel affine projection for nonlinearity tolerant optical short reach systems. *IEEE Trans. Commun.* **2020**, *68*, 6403–6412. [CrossRef]
37. Spyropoulou, M.; Kanakis, G.; Jiao, Y.; Stabile, R.; Calabretta, N.; Williams, K.; Dupuy, J.-Y.; Konczykowska, A.; Schatz, R.; Ozolins, O.; et al. Towards 1.6T datacentre interconnect technologies: The TWILIGHT perspective. *J. Phys. Photonics* **2020**, *2*, 041002. [CrossRef]
38. Zhou, X.; Urata, R.; Liu, H. Beyond 1 Tb/s intra-data center interconnect technology: IM-DD OR coherent? *J. Lightwave Technol.* **2020**, *38*, 475–484. [CrossRef]
39. Chacinski, M.; Westergren, U.; Stoltz, B.; Thylen, L.; Schatz, R.; Hammerfeldt, S. Monolithically integrated 100 GHz DFB-TWEAM. *J. Lightwave Technol.* **2009**, *27*, 3410–3415. [CrossRef]

40. Ozolins, O.; Pang, X.; Olmedo, M.I.; Kakkar, A.; Udalcovs, A.; Gaiarin, S.; Navarro, J.R.; Engenhardt, K.M.; Asyngier, T.; Schatz, R.; et al. 100 GHz externally modulated laser for optical interconnects. *J. Lightwave Technol.* **2017**, *35*, 1174–1179. [CrossRef]
41. Matsui, Y.; Schatz, R.; Che, D.; Khan, F.; Kwakernaak, M.; Sudo, T. Low-chirp isolator-free 65-GHz-bandwidth directly modulated lasers. *Nat. Photonics* **2021**, *15*, 59–63. [CrossRef]
42. Yamaoka, S.; Diamantopoulos, N.-P.; Nishi, H.; Nakao, R.; Fujii, T.; Takeda, K.; Hiraki, T.; Tsurugaya, T.; Kanazawa, S.; Tanobe, H.; et al. Directly modulated membrane lasers with 108 GHz bandwidth on a high-thermal-conductivity silicon carbide substrate. *Nat. Photonics* **2021**, *15*, 28–35. [CrossRef]
43. Chen, X.; Chandrasekhar, S.; Cho, J.; Winzer, P. Single-wavelength and single-photodiode entropy-loaded 554-Gb/s transmission over 22-km SMF. In Proceedings of the 2019 Optical Fiber Communications Conference and Exhibition (OFC), San Diego, CA, USA, 3–7 March 2019; pp. 1–3.
44. Yamazaki, H.; Nagatani, M.; Wakita, H.; Nakamura, M.; Kanazawa, S.; Ida, M.; Hashimoto, T.; Nosaka, H.; Miyamoto, Y. 160-GBd (320-Gb/s) PAM4 transmission using 97-GHz bandwidth analog multiplexer. *IEEE Photonics Technol. Lett.* **2018**, *30*, 1749–1751. [CrossRef]
45. Nagatani, M.; Wakita, H.; Jyo, T.; Mutoh, M.; Ida, M.; Voinigescu, S.P.; Nosaka, H. A 256-Gbps PAM-4 signal generator IC in 0.25 μm InP DHBT technology. In Proceedings of the 2018 IEEE BiCMOS and Compound Semiconductor Integrated Circuits and Technology Symposium (BCICTS), San Diego, CA, USA, 15–17 October 2018; pp. 28–31.
46. Baeuerle, B.; Heni, W.; Hoessbacher, C.; Fedoryshyn, Y.; Josten, A.; Haffner, C.; Watanabe, T.; Uhl, C.; Hettrich, H.; Elder, D.L.; et al. Reduced equalization needs of 100 GHz bandwidth plasmonic modulators. *J. Lightwave Technol.* **2019**, *37*, 2050–2057. [CrossRef]
47. Buchali, F.; Schuh, K.; Le, S.T.; Du, X.; Grözing, M.; Berroth, M. A SiGe HBT BiCMOS 1-to-4 ADC frontend supporting 100 GBaud PAM4 reception at 14 GHz digitizer bandwidth. In Proceedings of the 2019 Optical Fiber Communications Conference and Exhibition (OFC), San Diego, CA, USA, 3–7 March 2019; pp. 1–3.
48. Driad, R.; Makon, R.E.; Hurm, V.; Schneider, K.; Benkhelifa, F.; Losch, R.; Rosenzweig, J. In InP DHBT-based ICs for 100 Gbit/s data transmission. In Proceedings of the 2008 20th International Conference on Indium Phosphide and Related Materials, Versailles, France, 25–29 May 2008; pp. 1–4.
49. Jian, Y.; Pfister, H.D.; Narayanan, K.R.; Raghu, R.; Mazahreh, R. In Iterative hard-decision decoding of braided BCH codes for high-speed optical communication. In Proceedings of the 2013 IEEE Global Communications Conference (GLOBECOM), Atlanta, GA, USA, 9–13 December 2013; pp. 2376–2381.
50. Cvijetic, N.; Tanaka, A.; Ji, P.N.; Murakami, S.; Sethuraman, K.; Ting, W. First OpenFlow-based software-defined λ-flow architecture for flex-grid OFDMA mobile backhaul over passive optical networks with filterless direct detection ONUs. In Proceedings of the 2013 Optical Fiber Communication Conference and Exposition and the National Fiber Optic Engineers Conference (OFC/NFOEC), Anaheim, CA, USA, 17–21 March 2013; pp. 1–3.
51. GitHub. Available online: Github.com/dtu-dsp/Robochameleon (accessed on 2 February 2021).

MDPI
St. Alban-Anlage 66
4052 Basel
Switzerland
Tel. +41 61 683 77 34
Fax +41 61 302 89 18
www.mdpi.com

Applied Sciences Editorial Office
E-mail: applsci@mdpi.com
www.mdpi.com/journal/applsci

www.ingramcontent.com/pod-product-compliance
Lightning Source LLC
Chambersburg PA
CBHW062106220926
51850CB00010B/2910
* 9 7 8 3 0 3 6 5 4 9 4 7 7 *